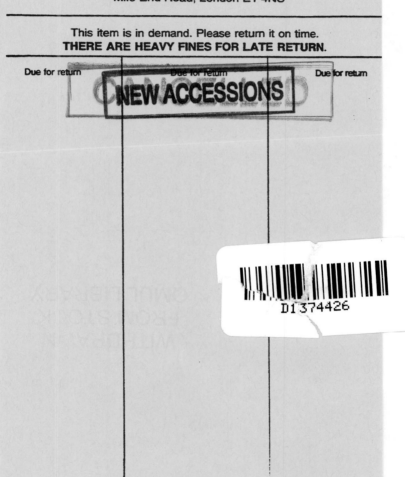

Quaternary Insects and Their Environments

Quaternary Insects and Their Environments

Scott A. Elias
Foreword by G. Russell Coope

SMITHSONIAN INSTITUTION PRESS
Washington and London

Editor and typesetter: Peter Strupp/Princeton Editorial Associates
Designer: Janice Wheeler

Library of Congress Cataloging-in-Publication Data
Elias, Scott A.
  Quaternary insects and their environments / Scott A. Elias.
    p.   cm.
  ISBN 1-56098-303-5 (cloth)
    1. Insects, Fossil.   2. Paleontology—Quaternary.   I. Title.
QE831.E45   1993
656′.7—dc20   93-18570
                                                CIP

British Library Cataloguing-in-Publication Data is available.

Manufactured in the United States of America
00 99 98 97 96 95 94   5 4 3 2 1

∞ The paper used in this publication meets the minimum requirements of the
American National Standard for Permanence of Paper for Printed Library Materials
Z39.48-1984.

For permission to reproduce illustrations appearing in this book, please correspond
directly with the owners of the works, as listed in the individual captions. The
Smithsonian Institution Press does not retain reproduction rights for these illustrations
individually or maintain a file of addresses for photo sources.

CONTENTS

FOREWORD

When Scott Elias asked me to write a few words of introduction to his book on Quaternary entomology, I was delighted—not because of any altruistic motive, but from a sense of relief that I had been absolved from the necessity of writing such a book myself. I have been waiting for a long time for a gap to open up in my research schedule, but, unlike generals who retire in middle age and write their memoirs, scientists are not in the habit of withdrawing from the fray so early, and this scientist has no intention of bowing out just yet. What a pleasure, therefore, to find a colleague who is both magnanimous enough to take on the task of writing this review and sufficiently diligent to search through the scattered literature and bring together a synthesis of a subject that straddles the no-man's-land between the sciences of biology, biogeography, geology, environmental science, and archaeology.

There can be little doubt that it is this interdisciplinary position that has been responsible for the neglect of these geologically recent insect remains. Biologists in general believed that fragmentary insects were almost impossible to identify and only very rarely identifiable to the species level. Furthermore, given the widely held belief that insects were evolving rapidly at the present day, Quaternary fossils represented extinct species or, at least the remote precursors of those that are still living today. Thus they were entitled to new and often fanciful species names, those given by Samuel Scudder to Late Pleistocene insects from Scarboro Bluffs, Ontario: such as, for instance *fossilis, dilapidatus, damnosum,* and *infernalis,* arranged here on an increasing scale of exasperation. Quaternary entomology was thus marginalized as an unimportant and therefore uninteresting sideline to the overall evolutionary narrative. Geologists likewise ignored these insect fossils, but for rather different reasons. They were so modern—"not even fossilized"—and for this reason fell outside the normal boundaries of palaeontology. They had little stratigraphic value, and the geologists already had such a wealth of pollen data as environmental indicators of glacial/interglacial cycles. What more could they possibly need? There seemed therefore to be very good reasons for relegating to the lowest priority level this rather esoteric brand of Quaternary science.

However, it is precisely these neglected areas of science that are likely to yield the most unexpected and therefore the most interesting results. Who would have anticipated that the study of Quaternary insect fossils would reveal that, for the most part, species have remained constant for hundreds of thousands of years (generations) and, in some cases, for millions of years. Evolutionary change is relatively easy to understand compared with this long-term species constancy, maintained in the face of the great climatic changes of the glacial/interglacial cycles. Stasis constitutes a much tougher and more interesting challenge to our understanding. The Quaternary fossil evidence, the only objective evidence that we have of what actually went on in the recent past, shows that insect species responded to these climatic changes by *moving* out of trouble rather than by *evolving* out of trouble. They tracked areas of acceptable climate as they shifted over the surface of the earth. In some cases species changed their geographical distributions by thousands of kilometers even during the last glacial period. Biogeographical history now becomes a science of testable hypotheses, since we can now reconstruct these changes of range within the context of a timed sequence of environmental events.

Far from being of little geological value, Quaternary insect fossils are proving to be sensitive indicators of past environments and climates. In particular they respond to changes of temperature more promptly and with greater intensity than any other component of the terrestrial flora and fauna. They are providing evidence that major climatic changes took place with quite unexpected suddenness, moving from glacial cold to interglacial warmth in a matter of decades rather than in millennia.

If all this is unfamiliar to you, let me recommend that you read this book, since it contains still more surprises about the geologically recent past in which our present-day insect species and communities were schooled. On the other hand, if you are already familiar with the subject, let me likewise encourage you to read this book. We Quaternary entomologists are few in number and thinly distributed in a hostile environment. We have much to learn from one another. I, for one, have thoroughly enjoyed reading this book, renewing old friendships and making new acquaintances within its pages.

G. Russell Coope
Royal Holloway College
· University of London

PREFACE

The field of Quaternary entomology has grown in the last three decades from a minor aspect of paleontology in the last century to one of the important disciplines within modern Quaternary studies. Enough organic detritus has been examined to show that insect exoskeletons are ubiquitous in many types of organic deposits dating back more than a million years. The more than 450 studies published in this field during the last 30 years show that insects are sensitive, reliable indicators of Quaternary environments. The stability of beetle species over hundreds of thousands or even millions of years has been established on the basis of abundant fossil evidence, and studies indicate that some insect species have shifted their distributions over vast regions in response to changing climates.

In spite of the hundreds of publications contributing to the body of knowledge in this field, the mass of data accumulated by paleoentomologists remains largely unknown to most neontologists. Some of the blame for this must be placed on the Quaternary entomologists themselves, for publishing most of their papers in geological journals and for giving most of their public lectures to geological audiences.

Another problem is that Quaternary entomology is perceived by some experimental biologists as "nineteenth-century science"; the description of fossil faunas has been equated by some with stamp collecting. However, as Morgan and Morgan (1987) pointed out, the days of the natural historian were not over at the turn of the twentieth century, and gathering "descriptive" data on species is more important now than ever, given the pace of natural habitat destruction around the world (Miller, 1991).

As Quaternary science grows and expands into many new lines of research, it becomes all too easy to be caught up in paleoenvironmental reconstructions based on one type of proxy data (e.g., pollen, insects, vertebrate remains, oxygen isotopes). Fossil insect research is still becoming known and accepted by colleagues in both neontology and paleontology. As with other newly developing disciplines, full acceptance will take time. Almost every audience I address requires a lengthy introduction to the subject of Quaternary entomology before I can get on with a discussion of my own research topic. One of the aims of this book is to provide a broad spectrum of readers with the necessary background information, in the hope that Quaternary entomology will be more widely understood.

ACKNOWLEDGMENTS

This book grew out of a seminar course in Quaternary entomology, that I taught at the University of Alaska, Fairbanks, during the fall of 1991. I thank my UAF colleagues Mary Edwards and David Hopkins for their support and encouragement, without which the book project would not have gotten off the ground.

I would like to thank my Quaternary entomology colleagues for their many contributions to this book. Robert Angus, Allan Ashworth, Ingolf Askevold, Daniel Berman, Russell Coope, and Donald Schwert provided figures or gave permission to use previously published figures. Paul Buckland, Russell Coope, and John Matthews served as manuscript reviewers. Matthews and an anonymous colleague went through the entire draft manuscript with a fine-tooth comb, finding many errors and instances of fuzzy thinking. Buckland and Coope (1991) authored a Quaternary entomology literature review that made my task much less tedious. Susan Short and John Hollin, as well as other colleagues at the Institute of Arctic and Alpine Research, University of Colorado, provided many useful comments and helped track down obscure site localities from various atlases and other publications. Kathleen Salzberg, managing editor of *Arctic and Alpine Research,* was a constant source of help in solving reference citation problems. Paul Carrara, of the U.S. Geological Survey, Denver, helped prepare all the scanning electron micrographs credited to me in the book. Peter Wigand and Saxon Sharpe, Desert Research Institute, University of Nevada, Reno, provided photographs of packrat middens.

I thank Peter Cannell, my editor at the Smithsonian Institution Press, for his much-needed help, support, and encouragement. Last but not least, I thank my family for putting up with my long hours, working weekends, and tirades throughout 1992.

Financial support for manuscript preparation was provided by a grant from the National Science Foundation for Long Term Ecological Research (BSR-9011658).

Quaternary Insects and Their Environments

1

THE HISTORY OF QUATERNARY
INSECT STUDIES

> The great resemblance of the insects to those now living, in most cases
> amounting to identity, shows that it takes a long time to effect a change in the
> Coleoptera.
> —Fordyce Grinnell (1908)

The study of Quaternary insect fossils has expanded rapidly during the last four
decades, and Quaternary entomology is achieving considerable success in
unraveling the history of changing environments and biotic responses to waxing and waning continental ice sheets. Unlike most other classical paleontological endeavors, Quaternary insect studies deal with the fossil remains of
species that still exist. Thus, the process of identification of Quaternary insects
primarily involves matching fossils to modern species—an approach that provides unusually accurate information on past environments because study of
the living populations of the insect species found in Quaternary assemblages
provides the necessary information on environmental tolerances, behavior, and
distribution patterns. Thus the work has provided a treasure trove of information on the shifting environments of the Quaternary.

Conversely, the study of Quaternary insect fossils provides new insights into
the corresponding modern fauna. For the first time, we now have data that
enable us to begin to answer important questions about species longevity,
centers of origin, and the stability of insect communities, questions that had
heretofore been considered only by theoreticians working back from modern
data. A body of hard fossil data is now at hand, and it sometimes conflicts
sharply with previous theories.

The principal aim of this book is to summarize and synthesize this fossil
evidence, so that neontologists, paleontologists, and others may understand its
importance and employ it in their own fields of study. Communication between
paleontologists and those who study modern biology has been limited. In an
age of specialization within these disciplines, few are capable of finding and
digesting all the necessary information in their own fields, much less the
interdisciplinary aspects of other fields. This book is intended to highlight

some of the important, if up to now dimly perceived, connections between Quaternary and modern entomology.

I shall focus on the Quaternary Period, with the exception of Chapter 11, in which information on some late Tertiary insects receives attention. Most insect fossil studies to date have dealt with late Quaternary faunal assemblages (i.e., those younger than 500,000 years), because younger deposits are more plentiful, and because most of the insect fossils they contain are in a better state of preservation than those found in deposits of early Quaternary age. In this chapter, I begin with an overview of the development of Quaternary entomology. Chapters 2–6 deal with methods used in the investigation of fossil insects, and with the value of insects in paleoecology, paleoclimatology, and zoogeography. Chapter 7 treats insect fossils from archeological sites. Chapters 8–11 provide summaries of studies from Europe, Asia, and the Americas. Chapter 12 offers conclusions and a prospectus for the future of the discipline.

Although the field of Quaternary entomology has dealt with insects from many orders, most studies have focused on the remains of the hard-bodied (highly sclerotized) groups, especially beetles (order Coleoptera). Beetles therefore take center stage in most of the faunal assemblages discussed in this book, although, as explained in Chapter 3, other insects are playing an increasingly important part in paleoenvironmental reconstructions.

A small but growing cadre of scientists is applying Quaternary insect studies to new regions and various types of organic deposits. Many of these researchers have been trained by G. Russell Coope at the University of Birmingham, or by his former students. However, these fruitful recent studies were preceded by almost a century of false starts and misconceptions. It is necessary to step back and examine this history, if only to gain a better appreciation of recent advances.

THE BEGINNINGS OF QUATERNARY ENTOMOLOGY

The intensive, systematic studies summarized in this book trace their lineage to an early series of intermittent, isolated studies in Europe and North America. These efforts were not very useful to the developing science of paleoecology because for the most part they were based on a false premise. The early investigators assumed that all fossil specimens, even those from late Quaternary deposits, represented extinct species. As we will see, this idea was pervasive not only during the nineteenth century, but also well into the twentieth. Armed with this notion of Pleistocene extinctions, a few workers began describing Quaternary insect fossils, applying such evocative names as *Helophorus pleistocenicus* (Lomnicki, 1894), *Olophrum interglacialie* (Mjöberg, 1904), and *Platynus exterminatus* and *Lathrobium antiquatum* (Scudder, 1900). Unfortunately, much of the fossil material described in these early papers has been lost, but many of the remaining specimens have subsequently been reidentified as belonging to extant species. A sampling of these reinterpreted fossils appears in Table 1.1.

Revisions of misidentified material by modern workers merit our attention because they show the extent and nature of the errors made by the first workers. The first is Angus's (1973) study of *Helophorus* (Hydrophilidae) fossils described by A. M. Lomnicki from Pleistocene deposits from Starunia and Borislav in the Ukraine. Lomnicki placed the *Helophorus* fossils from the Borislav site in five species, all described as new. Angus examined these fossils in Lomnicki's collection at Lvov and, because of their excellent state of preservation (Fig. 1.1), was able to perform a detailed study, including examination of the male genitalia of several specimens. All of the fossils were found to represent extant species (Table 1.1).

In 1877, Samuel H. Scudder published his first paper about insect fossils from the late Quaternary deposits at Scarborough, Ontario. During the next twenty years, Scudder described hundreds of beetle sclerites from the Scarborough Bluffs as representing extinct Pleistocene species. Scudder's work on the late Pleistocene insects from the Scarborough bluffs (Fig. 1.2) and elsewhere was an important, albeit flawed, first step for North American studies.

Scudder enjoyed dual careers in paleontology and entomology. He served many years as paleontologist to the U.S. Geological Survey, but also wrote lengthy monographs on the modern Lepidoptera of eastern North America. Most of his work on insect fossils dealt with Tertiary specimens. All told, he described 1144 insect species based on fossil material. Among these are 54 species of Pleistocene beetles. All but two were named as extinct species. Scudder had a reputation as an "excessive splitter" of genera (Cockerell, 1911, p. 341), and his contemporaries observed that he described too many species based on specimens that were preserved inadequately for satisfactory classification. Nevertheless, his contribution to paleoentomology was likened to the vertebrate paleontological contributions of "Leidy, Cope and Marsh combined" (Cockerell, 1911, p. 341).

One gains a better understanding of Scudder's difficulties in dealing with isolated fossil sclerites by noting some of the names he gave to them. For instance, the ground beetles he described include *Bembidion fragmentum, B. expletum,* and *B. damnosum; Pterostichus destitutus, P. depletus,* and *P. destructus;* and *Platynus dissipatus* and *P. dilapidatus* (Fig. 1.3).

Table 1.1

Beetle species described as extinct and subsequent extant identifications

"Extinct" taxon	Locality	Reference	=Extant species	Reference
Carabidae				
Elaphrus clairvillei lynni Pierce	Lynn Creek, British Columbia	Pierce (1948)	E. clairvillei Kby.	Goulet (1983)
Elaphrus irregularis Scudder	Scarborough, Ontario	Scudder (1890)	E. parviceps VD or E. americanus Dej.	Goulet (1983)
Elaphrus ruscarius foveatus Pierce	McKittrick, California	Pierce (1948)	E. finitimus Csy.	Goulet (1983)
Patrobus gelatus Scudder	Scarborough, Ontario	Scudder (1890)	P. cf. stygicus Chd.	Darlington (1938)
Patrobus decessus Scudder	Scarborough and Toronto, Ontario	Scudder (1900)	P. cf. stygicus Chd.	Darlington (1938)
Patrobus frigidus Scudder	Toronto, Ontario	Scudder (1900)	P. cf. stygicus Chd.	Darlington (1938)
Hydrophilidae				
Helophorus pleistocenicus Lomn.	Borislav, Ukraine	Lomnicki (1894)	H. sibiricus Angus	Angus (1973)
Helophorus dzieduszczkii Lomn.	Borislav, Ukraine	Lomnicki (1894)	H. aquaticus L.	Angus (1973)
Helophorus kuwerti Lomn.	Borislav, Ukraine	Lomnicki (1894)	H. oblongus LeC.	Angus (1973)

	Locality	Reference	Current name	Reference
Silphidae				
Nicrophorus guttula labreae Pierce	Rancho La Brea, California	Pierce (1949)	*N. marginatus* Fab.	Miller and Peck (1979)
Nicrophorus obtusiscutellum Pierce	Ranch La Brea, California	Pierce (1949)	*N. marginatus* Fab.	Miller and Peck (1979)
Nicrophorus mckittricki Pierce	McKittrick, California	Pierce (1949)	*N. marginatus* Fab.	Miller and Peck (1979)
Nicrophorus investigator alpha Pierce	Rancho La Brea, California	Pierce (1949)	*N. nigrita* Mannh.	Miller and Peck (1979)
Tenebrionidae				
Apsena labreae Pierce	Rancho La Brea, California	Pierce (1954a)	*A. laticornis* Csy.	Doyen and Miller (1980)
Coniontis blissi Pierce	Rancho La Brea, California	Pierce (1954b)	*C. abdominalis* LeC.	Doyen and Miller (1980)
Coniontis tristis alpha Pierce	Rancho La Brea, California	Pierce (1954b)	*C. abdominalis* LeC.	Doyen and Miller (1980)
Eleodes elongatus Grinnell	Rancho La Brea, California	Grinnell (1908)	*E. grandicollis* LeC.	Doyen and Miller (1980)
Eleodes behri Grinnell	Rancho La Brea, California	Grinnell (1908)	*E. osculans* (LeC.)	Doyen and Miller (1980)
Scarabaeidae				
Canthon simplex antiquus Pierce	Rancho La Brea, California	Pierce (1946)	*C. simplex* LeC.	Miller et al. (1981)

Figure 1.1. Modern (left) and fossil (right) specimens of *Helophorus oblongus,* misidentified by Lomnicki as *H. kuwerti.* (Photographs courtesy Robert Angus.)

Figure 1.2. Scarborough Bluffs, Toronto, Ontario.

Figure 1.3. Fossil beetles identified by Scudder from the Scarborough Formation, illustrated by Henry Blake in Scudder (1900). 1, *Bembidion expletum;* 2, *Badister antecursor;* 3, *Pterostichus depletus;* 4, *Patrobus decessus;* 5, *Bembidion damnosum;* 6, *Patrobus frigidus.* (Courtesy Geological Survey of Canada, Department of Energy, Mines and Resources. Reproduced with the permission of the Minister of Supply and Services Canada.)

Unfortunately, much of Scudder's fossil material has been lost. But the specimens that have been reexamined by modern taxonomists (Table 1.1) have been identified as belonging to extant species (Darlington, 1938; Goulet, 1983). Moreover, recent studies of insect fossils from modern exposures from the Scarborough Bluffs have yielded only specimens attributable to modern species (Morgan, 1972, 1975; Morgan and Morgan, 1976; Williams et al., 1981).

Carl Lindroth played an important role in establishing the modern aspects of the discipline of Quaternary entomology by his revision of previous Swedish studies of interglacial insects from Frösön, Härnön, Hälsingland, Pilgrimstad, and Angermanland (Lindroth, 1948). In particular, Eric Mjöberg had identified a number of fossil beetle sclerites from the Härnön site as representing extinct species (Mjöberg, 1904, 1905, 1915, 1916). Lindroth reviewed the Härnön material and found that "Mjöberg's opinion was extremely weakly founded" (1948, p. 4). Some of Mjöberg's extinct species represented extant taxa, and others had been placed in the wrong genera. Thus Lindroth concluded that all of Mjöberg's species names from the Härnön site were junior synonyms of extant species. Mjöberg had attempted to identify various sclerites, such as staphylinid elytra, to the species level—an impossible task for most specimens. He was also faced with many postmortem changes in his specimens, which led to more difficulties. Lindroth summarized Mjöberg's problems as follows:

It is not difficult to understand how the Härnön material could make so strange an impression on Mjöberg that he suggested it to be the remains of a partly extinct fauna. He had met with the manifestation of a fact, the difficulty of which could be expressed in the form of the following question: To what extent do fossils and subfossils change by postmortem processes?

Postmortem changes in fossil insect sclerites had already been the subject of some debate in Scandinavia (Henriksen, 1933). Lindroth noted that the normal punctulae on the exoskeletons of ground beetles (Carabidae), rove beetles (Staphylinidae), and pill beetles (Byrrhidae) were deepened in the interglacial fossil specimens. He tried to reproduce these changes by treating modern specimens with several acids and bases, but could not achieve the same effects.

Lindroth reported several other important advances in this paper, including the observation that, in dealing with fossil material, the most useful taxonomic characters are those in the cuticular microsculpture (the microscopic lines and meshes formed on the surface of the exoskeleton). This character system has been relied upon for identification of fossil specimens ever since (Coope, 1970, 1986). Lindroth identified 42 species of Hemiptera and Coleoptera from the interglacial assemblages and made paleoenvironmental reconstructions based on the species' modern distributions and ecological requirements.

Strobel and Pigorni (1864) published one of the first studies to attempt a paleoenvironmental reconstruction based on fossil insect species. Their work dealt with an archeological site in Italy. Other early European studies were mostly paleontological in focus. These included work in France (Fliche, 1875, 1876), Germany (Flach, 1884; Schaff, 1892; Kolbe, 1894), Switzerland (Heer, 1865), Denmark (Wesenberg-Lund, 1896), and Finland (Andersson, 1898). British studies began with Bolton (1862) and Wollaston (1863). Shotton et al. (1962) reexamined Wollaston's specimens, and found them to represent extant taxa. The practice of naming extinct species of beetles from Pleistocene assemblages was continued through the first half of this century. Coope (1968a) discussed this problem in his reinterpretation of Lesne's unpublished (circa 1920) Pleistocene "arctic" fauna from Barnwell Station, Cambridge.

Nineteenth-century studies in North America have been summarized by Ashworth (1979). They include work in the eastern United States by Horn (1876), at other Canadian and eastern United States sites by Scudder (in Ami, 1894; Scudder, 1898), and in Illinois by Wickham (1917). All of these papers primarily describe fossils as extinct Pleistocene species. However, in 1919, Wickham published a paper on fossil insects from Vero Beach, Florida, in which he stated that the fossils probably represented extant species, a hypothesis later supported by Young (1959).

The final chapter in this history of first attempts was written on the famous asphalt deposits of southern California and comprises a series of studies that have been summarized by Miller (1983). In 1908 Grinnell began to study fossil insects at the Rosemary site, near Los Angeles. He was perhaps the first worker to abandon the assumption that all Pleistocene fossils must represent extinct forms, and made matches of the fossil material with many modern species (Ashworth, 1979). Pierce began publishing his extensive work on the Rancho La Brea and McKittrick sites in 1944, and continued for two decades (Pierce, 1946, 1948, 1949, 1954a,b, 1957). Although Pierce appreciated the significance of Pleistocene insects as paleoenvironmental indicators, his taxonomic work is fraught with errors. He continued the practice of naming species and subspecies of "extinct" beetles. Most of his fossil material has been reexamined recently, and nearly all of the specimens have been matched with extant species (Miller and Peck, 1979; Doyen and Miller, 1980; Miller et al., 1981; Goulet, 1983; Miller, 1983; Wilson, 1986). However, Miller et al. (1981) could not find modern species to match two scarab taxa described by Pierce, *Onthophagus everestae* and *Copris pristinus*. The species most closely related to these in modern collections live in Texas and Mexico and feed on mammal dung. Miller (1983) has speculated that the extinction of many mammalian species at the end of the Pleistocene would have caused a reduction in dung

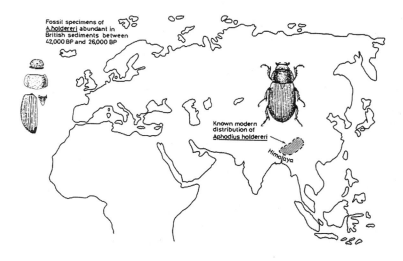

Figure 1.4. Fossil and modern localities for *Aphodius holdereri*. (After Coope, 1973.)

availability for these beetles, leading to their extinction. Increasing aridity also may have played a role in the beetles' demise. However, Miller noted that these species may yet be found living today in inadequately studied regions of Mexico.

This last possibility was demonstrated by Coope, in his search for a modern counterpart to a fossil dung beetle species from the British Pleistocene. Specimens of this *Aphodius* were common enough in deposits from the middle of the last glaciation, but for years Coope failed to find a modern species that matched it. Finally, he searched through uncatalogued beetle collections from the 1924 British Everest Expedition, housed in the British Museum, and found several unnamed specimens that were a perfect match for the fossils (Coope, 1973). The modern specimens were determined to be *A. holdereri*, a species known today only from the Tibetan Plateau (Fig. 1.4). The correspondance of fossils to modern specimens was confirmed through the comparison of male genitalia, considered among the most diagnostic of the sclerotized body parts of beetles.

COOPE'S PIONEERING WORK IN BRITAIN

In 1955, Russell Coope began studying Quaternary insect fossils at the Upton Warren site, near Birmingham. When Coope began this work, he had little knowledge of paleoentomology. He first visited the Upton Warren site on behalf of the geological museum in his department to collect Pleistocene mammal bones protruding from the exposure. However, an examination of the organic sediments surrounding the bones revealed numerous, shiny fragments of beetles. These fossils were initially in an excellent state of preservation and Coope decided to try to identify them. By his own admission, he was "blissfully unaware" of the difficulties he would encounter in dealing with taxonomists who had been taught that all Pleistocene insect fossils must represent extinct species (Coope, 1965). By making patient comparisons with modern specimens in the natural history collections of the Birmingham City Museum, Coope matched most of the Upton Warren material with modern species, and gradually became convinced that all of the fossil material could eventually be identified as belonging to species included in the modern fauna. His refusal to accept the conventional dogma of Pleistocene extinction and his keen interest in the fossils led him to persevere in their study until he managed to win over many of his paleontological and modern entomological colleagues. This first endeavor, and his first paper on the subject (concerning the Chelford site) in 1959, owed much to the support of Coope's departmental head in the Geology Department at Birmingham, Professor Fred Shotton, who provided "constant encouragement . . . boundless enthusiasm, and . . . practical assistance during the course of the work" (Coope, 1959, p. 85). The Upton Warren paper was published a few years later (Coope et al., 1961).

Coope is a dedicated naturalist with wide-ranging interests and a driving curiosity. His determination, often in the face of considerable opposition from entomologists and Quaternary scientists alike, brought the field of Quaternary entomology out of obscurity and into the limelight.

Since publishing his seminal works, Coope has gone on to study assemblages throughout the British Isles and western Europe, publishing more than a hundred papers on Quaternary insects (Buckland and Coope, 1991). His contributions to the field have been outstanding. He has established most of its fundamental principles and has trained dozens of students. Moreover, he is a gifted public speaker, and his great enthusiasm for Quaternary entomology has captured the imagination of audiences at geological and entomological meetings for more than three decades.

THE DEVELOPMENT OF QUATERNARY ENTOMOLOGY OUTSIDE BRITAIN

Coope and his colleagues Shotton and Peter Osborne at the University of Birmingham worked more or less in isolation through the 1950s and early

1960s. Since then, however, several of their students have established their own laboratories and research programs. As of this writing, 37 scientists in nine countries are studying Quaternary or Late Tertiary insect fossils. Their names, affiliations, and research interests are summarized in Table 1.2. Great Britain, the cradle of Quaternary insect studies, still has the greatest concentration of workers. Fred Shotton died in 1990, but Russell Coope and Peter Osborne remain in Birmingham. Paul Buckland is in Sheffield, Harry Kenward is in York, Robert Angus is in Egham, Surrey, and Bridget Wilkinson is in London. The majority of British workers are principally concerned with insect fossils from archaeological sites, although Wilkinson specializes in caddisfly larvae.

Continental Europe remains underrepresented in Quaternary entomology. Following his training at Birmingham, Philippe Ponel moved to Marseille, France. In Amsterdam, the Netherlands, Tom Hakbijl and Bas Van Geel are studying insects from archaeological sites. Jaap Schelvis, in Groningen, works on fossil mites. Alexander Klink, in Wageningen, studies aquatic insect fossils, as does Wolfgang Hofmann, in Plön, Germany. In Scandinavia, Jens Böcher (Copenhagen) is the only Danish researcher and Geoffrey Lemdahl (Lund) is the only Swede. Böcher is also investigating insect fossils of late Tertiary through Holocene age from Greenland.

Russian scientists in this field have worked in relative isolation from their western colleagues until recently. Nevertheless, they have developed extensive research programs, one of which is based in the laboratory of Vladimir Nazarov in Minsk in Byelorussia. Fedor Bidashko in Uralsk has concentrated his studies in the Ural region. Most of the other Russians have focused on the arctic and subarctic fossil faunas of Siberia. Vertebrate paleontologist Andre Sher in Moscow played an important role in getting the Siberian studies under way. He was able to obtain permission, at a time when such permission was seldom given, to send samples from the Kolyma region to John V. Matthews, Jr., in Ottawa, Ontario. Matthews's preliminary report stimulated further research by the Russian scientist Sergei Kiselyov, who went on to publish a monograph on the fossils from Quaternary deposits of the Kolyma lowland and other Siberian regions. Since the 1970s, additional studies have been performed by Daniel Berman in Magadan (Soviet Far East), and by Alexandr Druk, Kiselyov, and Dmytry Krivolutsky, in various institutions in Moscow. Krivolutsky and Druk study fossil mites.

North American studies of the modern era began in 1968 with Matthews's graduate studies of Quaternary insects from the Alaskan interior. Matthews has since joined the Geological Survey of Canada at Ottawa. Robert E. Nelson is working in Waterville, Maine. Scott Miller, now in Honolulu, Hawaii, has studied the asphalt deposit insects of California. Richard Morlan, also in

Table 1.2
Quaternary entomologists and their specialities

Name	Affiliation	Area(s) of interest
Robert Angus	Department of Zoology, Royal Holloway College, Egham, Surrey, United Kingdom	Palearctic Pleistocene *Helophorus*
Allan Ashworth	Geology Department, North Dakota State University, Fargo, North Dakota, USA	Late Quaternary insects of the American Upper Midwest and of southern South America
Valerie Behan-Pelletier	Biosystematics Research Centre, Agriculture Canada, Ottawa, Ontario, Canada	Oribatid mites, including Quaternary fossils
Daniel Berman	Institute of Biological Problems of the North, Magadan, Russia	Pleistocene insects of northern Siberia
Fedor Bidashko	Ural Anti-Pest Station, Uralsk, Russia	Pleistocene insects of the Ural region
Jens Böcher	Zoological Museum, Copenhagen, Denmark	Quaternary and late Tertiary insects of Greenland and Denmark
Paul Buckland	Department of Prehistory and Archaeology, University of Sheffield, Sheffield, United Kingdom	Insect fossils from archaeological sites in Britain, Iceland, and Greenland
G. Russell Coope	Geology Department, University of Birmingham, Birmingham, United Kingdom	Pleistocene insects of Britain and western Europe
Alexandr Druk	Severtzoff Institute of Animal Morphology and Ecology, Moscow, Russia	Oribatid mites of Siberia and Alaska
Scott A. Elias	Institute of Arctic and Alpine Research, University of Colorado, Boulder, Colorado, USA	Quaternary insects of Alaska, the Rocky Mountain region, and the Great Basin and Chihuahuan deserts; insects from archaeological sites in North America
Clarke Garry	University of Wisconsin, River Falls, Wisconsin	Late Quaternary insects of the Great Lakes region
Tom Hakbijl	Instituut voor Prae- en Protohistorische Archeologie, Amsterdam, The Netherlands	Insects from archaeological sites in The Netherlands
Wolfgang Hofmann	Max-Planck-Institut für Limnologie, Plön, Germany	Late Quaternary Chironomidae and Cladocera fossils from Germany and Switzerland

(continued)

Table 1.2 (*Continued*)

Name	Affiliation	Area(s) of interest
Harry Kenward	Unit for Environmental Research—Archaeology, York Archaeological Trust, York, United Kingdom	Insects from archaeological sites in Britain, Norway, and the Orkney Islands
Sergei Kiselyov	Department of Paleogeography, Moscow State University, Moscow, Russia	Quaternary insects of Siberia
Alexander Klink	Hydrobiological Consultants, Wageningen, The Netherlands	Late Quaternary aquatic insects of The Netherlands
Dmytry Krivolutsky	Severtzoff Institute of Animal Morphology and Ecology, Moscow, Russia	Fossil oribatid mites of Russia
Geoffrey Lemdahl	Kvartärbiologiska Laboratoriet, University of Lund, Lund, Sweden	Late Quaternary insects of Scandinavia, Poland, and Switzerland
John V. Matthews, Jr.	Geological Survey of Canada, Ottawa, Ontario, Canada	Late Tertiary insects of the high Arctic; Quaternary insects of Canada
Randall Miller	Geology Department, New Brunswick Museum, St. Johns, New Brunswick, Canada	Late Quaternary insects of Canada (especially the Maritime Provinces)
Scott Miller	Entomology Department, Bishop Museum, Honolulu, Hawaii, USA	Late Quaternary insects from the California asphalt deposits
Alan Morgan	Department of Earth Sciences, University of Waterloo, Waterloo, Ontario, Canada	Quaternary insects of eastern North America (especially the Great Lakes region)
Anne Morgan	Department of Biology, University of Waterloo, Waterloo, Ontario, Canada	Quaternary insects of eastern North America (especially the Great Lakes region)
Richard Morlan	Archeological Survey of Canada, National Museum of Man, Ottawa, Ontario, Canada	Quaternary insects of the Yukon Territory
Vladimir Nazarov	Institute of Geochemistry and Geophysics, Byelorussian Academy of Sciences, Minsk, Russia	Cenozoic insects of Byelorussia
Robert E. Nelson	Department of Geology, Colby College, Waterville, Maine, USA	Quaternary insects of Alaska and New England

Table 1.2 (*Continued*)

Name	Affiliation	Area(s) of interest
Carl Olson	Entomology Department, University of Arizona, Tucson, Arizona, USA	Late Quaternary insects of the Sonoran Desert (Mexico and Arizona)
Peter Osborne	Geology Department, University of Birmingham, Birmingham, United Kingdom	Insects from British archaeological sites
Jerry Pilny	Department of Earth Sciences, University of Waterloo, Waterloo, Ontario, Canada	Quaternary insects of Canada
Philippe Ponel	Laboratory of Paleobotany and Palynology, URA, CNRS, Marseille, France	Quaternary insects of France
Jaap Schelvis	Biologisch-Archaeologisch Instituut, Rijksuniversiteit Groningen, The Netherlands	Fossil mites from archaeological sites in The Netherlands
Donald Schwert	Geology Department, North Dakota State University, Fargo, North Dakota, USA	Late Quaternary insects from the upper Midwest of the United States; biogeographic origins of boreal insect faunas
Andre Sher	Severtsov Institute of Evolutionary Animal Morphology and Ecology, Russian Academy of Sciences, Moscow, Russia	Quaternary insects of Siberia
Bas Van Geel	Hugo de Vries–Laboratorium, University of Amsterdam, Amsterdam, The Netherlands	Insects from archaeological sites in The Netherlands
Ian Walker	Department of Biology, Okanagan College, Kelowna, British Columbia, Canada	Quaternary Chironomidae of Canada
Bridget Wilkinson	15 Manor Drive, Southgate, London, United Kingdom	Quaternary caddisfly larvae of Europe
Nancy Williams	Life Sciences Division, Scarborough College, University of Toronto, West Hill, Ontario, Canada	Quaternary caddisfly larvae of North America and Europe

Ottawa, is an archaeologist who occassionally works on insect fossils. Carl Olson, in Tucson, Arizona, collaborates with paleobotanist Thomas Van Devender on studies of Sonoran Desert insect fossils. Ian Walker, in Kelowna, British Columbia, investigates aquatic insect fossils. Valerie Behan-Pelletier (Ottawa, Ontario) specializes in oribatid mites (including fossils), and Nancy Williams (West Hill, Ontario) in fossil caddisfly larvae.

Three of Coope's former students have developed Quaternary entomology programs in North America: Allan Ashworth (Fargo, North Dakota), Anne Morgan (Waterloo, Ontario), and I. Alan and Anne Morgan, in turn, have supervised the graduate theses of Donald Schwert (now in Fargo), Randall Miller (now in St. John, New Brunswick), and Jerry Pilny (who remains in Waterloo). Schwert helped to train Clarke Garry, who now works in River Falls, Wisconsin. I am currently working in Boulder, Colorado. All of us focus mainly on beetle remains.

2

METHODS

The main problem when dealing with Quaternary insect fossils is embarrassment of riches.
    —Russell Coope (1986)

In this chapter, I provide answers to the questions about methodology most frequently asked by geologists, archaeologists, students of Quaternary entomology, and other biologists.

FOSSIL PRESERVATION

The remains of robust insects (especially beetles but also including ants, caddisfly larvae, bugs, and various genera within other orders) are abundant in unconsolidated organic deposits of Quaternary age. In the arctic, this type of preservation extends back several million years in permanently frozen sediments of late Tertiary age (Matthews, 1977a). The remains found in these sediments are actual exoskeletons, not mineral replacements or impressions in fine-grained sediments. Insect tracks, or *lebensspuren,* are also found occasionally in Holocene muds (Ratcliffe and Fagerstrom, 1980).

The remarkable preservation of insect exoskeletons in Quaternary sediments is due in large part to the skeletons' composition. Insect body walls are made up of sclerites: individual plates separated by sutures. These in turn are composed of two principal layers: a thin, more or less waxy, outer epicuticle that serves to prevent water loss, and a thicker procuticle just beneath (Fig. 2.1). The procuticle is composed of an outer layer (the exocuticle) and an inner layer (the endocuticle). The exocuticle contains a tough, durable substance called sclerotin. This forms during the initial hardening of the exoskeleton of the emerging adult, when cuticular proteins are exposed to secreted quinones that cause hardening by the formation of cross-links between protein molecules. One reason that beetles are the dominant group in most Quaternary fossil assemblages is that many beetle exoskeletons are very heavily sclerotized—a

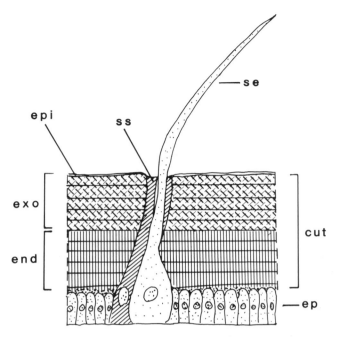

Figure 2.1. Cross section of insect cuticle (after Borror et al., 1981). cut, Cuticle; end, endocuticle; ep, epidermis; epi, epicuticle; exo, exocuticle; se, seta; ss, setal socket.

fact readily apparent to anyone who has attempted to put an entomological pin through modern specimens of weevils (Curculionidae), darkling beetles (Tenebrionidae), and ground beetles (Carabidae).

The procuticle is made chiefly of chitin, a nitrogenous polysaccharide of formula $(C_8H_{13}NO_5)_n$. Chitin is a very resistant substance that is insoluble in water, alcohol, and dilute acids and bases (Borror et al., 1981). Furthermore, it is not broken down by the digestive enzymes of mammals, and observation of mammal scat will reveal many intact insect parts (see Elias and Halfpenny, 1991).

TYPES OF SEDIMENTS CONTAINING INSECT FOSSILS

Insect exoskeletons are found chiefly in anoxic sediments that contain abundant organic detritus. Insects decompose rapidly in heavily oxidized sediments, leaving either thin, partially preserved sclerites or no trace at all. Water-lain sediments are generally the best source of insect fossils, because water acts to concentrate the insects. In contrast, ancient soils (paleosols) contain few fos-

Figure 2.2. Deltaic deposit at Lake Isabelle, Colorado. (Photograph courtesy Susan K. Short.)

sils, because insects that die on a soil surface are very widely dispersed across a landscape. The chemical processes involved in soil formation also cause degradation of insect fossils.

Lakes, ponds, and kettleholes serve as reservoirs that collect insects, and sediments that accumulate in these waters act rapidly to cover their remains, preventing oxidation. The best lake sediments for study are those that contain abundant organic detritus (plant macrofossils, insect exoskeletons, and other organic debris). This type of sediment usually accumulates in the shallow waters of the littoral zone. Sediments rich in organic detritus are often found where a small stream enters a lake, because the flotsam carried by the stream tends to settle out rapidly where the stream enters standing water (Fig. 2.2) (see Elias, 1985, p. 31).

Fluvial sediments often accumulate detritus, especially in secondary channel bends, backflows, and pools between riffles. Oxbow lakes, formed when a

Figure 2.3. Organic horizon in coastal bluff exposure, Kvichak Peninsula, Alaska.
(Photograph by the author.)

bend in a large stream becomes cut off from the main channel, accumulate a
sequence of organic sediments representing the transition from fluvial to
lacustrine environments, often finishing with infilling by bog vegetation.

The aim in sampling these sediments is to obtain a sufficient quantity to
yield at least a liter of organic detritus. Generally, the quantity of sediment
obtainable by piston coring of lake and pond sediments is insufficient to yield
adequate amounts of detritus. Although some success has been achieved by
taking multiple, large-diameter cores, this approach is not often practical and
may be hampered by difficulties in correlating the horizons among the cores,
unless numerous marker horizons are present (see Hoganson and Ashworth,
1992, pp. 102–104). The best results are obtained by sampling exposures of
sediments, either natural or man-made. Natural exposures include cut banks of

Figure 2.4. Peat profile exposed in bluff at Ennadai Lake, Keewatin, Northwest Territories, Canada. (Photograph courtesy Harvey Nichols.)

streams and ocean and lake bluffs (Figs. 2.3 and 2.4). Man-made exposures include those in gravel and clay pits (Fig. 2.5), irrigation ditches, trenches, and building sites. At times, even man-made catastrophes may be put to good use. For instance, organic sediments deposited along the banks of the Roaring River in Colorado were exposed for the first time in 1982 when an earthen dam failed several kilometers upstream, resulting in a torrent of water that scoured the river channel (Elias et al., 1986). Similarly, lake sediments were exposed at Lake Emma in Colorado (Fig. 2.6) when a mine roof beneath the lake collapsed in 1978, catastrophically draining the lake (Elias et al., 1991).

In many of these exposures sediments rich in organic matter are made up of detritus dispersed in sand, silt, and clay. For insect fossil analysis, the best results are obtained from organic detritus in silts or fine sands. Not only does

Figure 2.5. Late Pleistocene (Scarborough Formation) organic sediments exposed in Don Valley Brick Pit, Toronto, Ontario. (Photograph by the author.)

the silt readily disaggregate and separate from the organics in water screening, but its limited compressibility ensures that the fossils are preserved in their original shape (Coope, 1986).

Small bogs and fens are rich sources of insect fossils. Mosses and sedges may accumulate so rapidly in shallow water that they do not decompose, but rather build up into layers of peat that may also trap insect exoskeletons (Figs. 2.7 and 2.8). Buckland (1976a) noted that fen peats tend to be richer in insect remains than acid peats (such as *Sphagnum* peats) because fen peats offer a greater diversity of insect habitats. Some peat bogs in northern countries may cover thousands of square kilometers. The centers of large bogs tend to be rather sterile in terms of insects, and the best samples of insect fossils are usually obtained from the edges of bogs. As with lake sediments, peat samples are taken most readily from exposures.

Peats overlain by glacial ice or by large volumes of other sediments may become compressed or felted, making insect fossil extraction difficult. Coope (1961) and his colleagues at the University of Birmingham first dealt with felted peats by splitting them along bedding planes and examining the resulting surfaces for fossils. This technique had the advantage of yielding some nearly intact specimens, found together on the bedding planes. However, the procedure tended to overemphasize large, brightly colored specimens and to miss the

Figure 2.6. Top: Lake Emma, Colorado, in 1921. (Photograph courtesy San Juan County Historical Society.) Bottom: Lake Emma basin in 1979 after catastrophic drainage of the lake into a mine beneath it the previous year. (Photograph courtesy Paul Carrara.)

Figure 2.7. Sedge peat exposure at La Poudre Pass, Rocky Mountain National Park, Colorado. (Photograph by the author.)

small, dull specimens that generally represent the majority of species in any given sample.

In the last ten years, Quaternary insect fossils have also been discovered in packrat (*Neotoma* spp.) middens in the American Southwest. Packrats are known to bring a variety of objects to their den sites, whether to use as food, out of curiosity, or for protection. As the objects—including edible plants, cactus spines, vertebrate and insect remains, small pebbles, and feces—accumulate, they are cemented into black tarry masses called middens by the packrats' urine (Fig. 2.9). As the urine dries, it hardens, cementing layers of midden material that dry in rockshelters, preserving a paleoecological record that may span thousands of years. Packrat middens are unlike water-lain deposits: physical factors (e.g., current speed, size of catchment basin, and facies) generally determine the composition of water-lain deposits, whereas the packrat is the most important agent of accumulation of the organic material preserved in middens. Modern studies have shown that the majority of insects associated with packrat nests in rock shelters and caves represent species that generally live outside such shelters, but come in to share the microenvironment created by the packrat. These insects, called facultative inquilines, either prey on other insects in and around the rat's nest or scavenge plant materials there (Elias, 1990a).

Figure 2.8. The author sampling a *Sphagnum* peat exposure on the Foraker River, Denali National Park, Alaska. (Photograph courtesy Susan K. Short.)

SAMPLING PROCEDURES

The most informative paleoenvironmental reconstruction of a given site will not be based on isolated study of any one type of fossil. Instead it will be the fruit of the combined efforts of a team of investigators collaborating in the study of a broad spectrum of fossils, including insects, pollen, plant macrofossils, diatoms, mollusks, and ostracodes. Stratigraphy, geomorphology, and geochronology also play a vital role (see the study of Lobsigensee, Switzerland, described in Ammann et al., 1983). Obviously, this type of collaborative effort requires much advance planning before field work is undertaken. For fossil insect work, as for these other disciplines, it is important to locate and

Figure 2.9. Top: Packrat midden in a rockshelter on Badger Mountain, Pahranagat Range, Lincoln County, Nevada. Bottom: Close-up view of a packrat midden sample from a rockshelter in Pahranagat Wash, Lincoln County, Nevada. (Photographs by Peter Wigand, courtesy Desert Research Institute, University of Nevada, Reno.)

sample exposures so as to allow the collection of other types of fossils, such as seeds, pollen, and material suitable for dating (Coope, 1986).

Before sampling an exposure for insect fossils, the investigator must cut into the exposed face to remove material that has been exposed to repeated wetting and drying, and to eliminate contamination by modern insects that burrow into exposed banks. This procedure will also help to ensure that potential radio-carbon samples will not be contaminated by modern carbon from rootlets, mold, mosses, and other recent organic material (Morgan, 1988).

Once the exposure has been cleaned, and a baseline stratum (an easily observable feature such as a distinct sand lens or a lens of volcanic ash) established, horizons to be sampled for insects are generally measured in 5- or 10-cm increments. It is useful to photograph and sketch the exposure before sampling, and some workers now take "instant" photographs of exposures with a Polaroid camera and include the photographs in their field notebooks along with verbal descriptions. A video camcorder is another useful tool in the field to document sites more fully.

Some discrete, organic-rich horizons in lake, pond, and fluvial sediments yield more insect fossils than bulk sediments with little organic content. It is preferable to sample these units in intervals of no more than 5 cm, so that the insect assemblages from a single sample represent a discrete time interval. However, in some cases, the organic-rich lenses in an exposure may be too few in number to document a complete sequence of changes in insect faunas through time.

When that is the case, less organic-rich sediments must be sampled, and large quantities of sediment must be sieved to obtain sufficient organic detritus to study. This is best done in the field using a bucket sieve with a 300-μm screen (Fig. 2.10). Sampling laterally for several meters along an exposure may yield tens or even hundreds of kilograms of silts and fine sands, of which 3–5% will be organic detritus.

It is possible to wash large volumes of fine-grained sediments (except clays) rapidly through the sieve screen near the base of the bucket, accumulating a quantity of organic detritus on the screen. This process requires a water source in close proximity to the exposure (not to mention a strong back). Ponds and rivers in which samples are sieved often contain entrained insect fragments, either modern or fossil, and these fragments may commingle with the sieved residue if water is allowed to wash in over the top of the bucket sieve.

The volume of sediment required to provide sufficient organic detritus varies from site to site and from horizon to horizon. For instance, to obtain 1 kg of organic detritus, Nelson (1982) removed about 1000 kg of sediments from one horizon in an exposure on the Ikpikpuk River on the North Slope of Alaska. In

Figure 2.10. Susan Short bucket-sieving organic sediments at a collecting site on Arctic Bay, Baffin Island, Northwest Territories, Canada. (Photograph courtesy Edward Rowan.)

contrast, only 20 kg of lake sediments from Lobsigensee, Switzerland, provided the requisite liter of organic detritus (Elias and Wilkinson, 1983).

Once the organic detritus samples have been collected (either through sampling of organic-rich horizons or through the concentration of organic matter by field sieving), each sample should be placed, still damp, in a heavy-gauge plastic bag and sealed shut. If the sample is allowed to dry, the insect sclerites contained in the sample tend to become curled and to split into small fragments. Each bag must be clearly labeled.

Peat exposures may be sampled in blocks from a cleaned face (Fig. 2.11). The blocks should be at least 10–15 cm across and 10–15 cm deep, and taken in vertical increments of 5 cm. However, in order to expedite the field work, larger blocks (representing depth intervals of 10, 15, or 20 cm) may be taken

Figure 2.11. Harvey Nichols sampling peat at the La Poudre Pass site, Rocky Mountain National Park, Colorado. (Photograph by the author.)

and then split into 5-cm depths at a later date in the laboratory. The peat blocks should be wrapped in heavy-gauge aluminum foil and labeled to show the orientation of the block (top, bottom, front, back) and the depth interval. The wrapped block should then be sealed in a plastic bag that is also clearly labeled.

In most cases, study of a site is a once-in-a-lifetime opportunity, either because of the high cost of travel or because of the ephemeral nature of the sampling site itself. Ditches tend to be filled in, stream-cut exposures slump or disappear after the stream changes course, frozen sediments thaw, and building sites are covered with concrete. In light of the unique opportunity presented by a given sampling site, redundancy in sampling is very important. An archive portion of every sample studied should be saved, whenever possible, for future study. When a sample is sieved in the field, it is also advisable to collect an unsieved sample for laboratory preparation, as a check against contaminants in the sieved sample. Redundancy of sampling also offers the possibility that, even if some of the samples are lost by baggage handlers or mail clerks, or are left in cold storage for a decade, the investigator will still be assured of a useful set of samples.

It is worth noting here that not all valuable insect fossil samples have come from researchers exploring well-developed exposures. For instance, I have begun working on insect fossils from terrestrial peats from near the top of marine sediment cores taken in the shallow waters of the Chukchi Sea, off the northwest coast of Alaska. The geologists who obtained these cores were more interested in the marine sediments they contained, but the terrestrial peats near the top of the cores contain insects that once lived on the exposed continental shelves of the Bering Land Bridge (Elias et al., 1992a). I have also retrieved useful insect assemblages from pond sediment scraped from the brain case of a mammoth in the Wasatch Mountains of Utah (Elias, 1990b) and from dung taken from the frozen carcass of a mammoth in Alaska (Elias, 1992a). Valuable information may come from insects extracted from samples taken in a variety of contexts.

The collection of insect fossil samples from archaeological sites is a some-what specialized process (Buckland, 1976a). It is vital that proper sampling methods be used, and that the fossil insect investigation be coordinated with the archaeological project as a whole. Sampling intervals from archaeological sites may need to be smaller than those used in natural deposits, because the paleoentomologist is often called upon to develop a paleoenvironmental re-construction of a discrete interval (e.g., a single human occupation horizon in a sequence). As in strictly paleontological studies, it is best for researchers to take their own samples, in order to develop a clear understanding of the site and its peculiarities.

EXTRACTION AND CONCENTRATION OF INSECT FOSSILS FROM SEDIMENTS

The sequence of steps in extraction and mounting of insect fossils is sum-marized in Figure 2.12. Fortunately, fossil insect extraction is a relatively safe, inexpensive, and easy process. No costly chemicals or equipment are required, and less time is needed than for the preparation of pollen or diatom samples, for instance. The only lengthy process that may be involved is the pretreatment of samples to disaggregate the organic detritus from the inorganic matrix (such as calcareous sediments and clays) or to soften and then disperse felted peats or

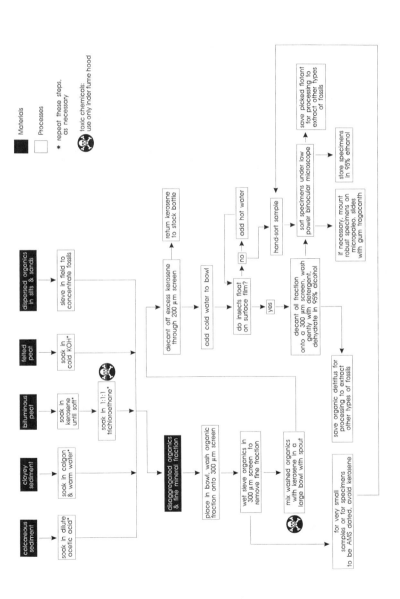

Figure 2.12. Summary of methods for extraction of insect fossils from various types of sediments. Darkened boxes represent materials; white boxes represent processes. Treatment of calcareous, clayey, and bituminous sediments and felted peats after Morgan (1988).

lignite. Such procedures may last hours or weeks, and may have to be repeated several times in order to be effective.

Processing procedures are largely a matter of personal preference and availability of equipment, and some workers use a slightly different procedure than the one discussed here. Once disaggregated organic detritus is obtained, the next step is to wet-screen the sample in a 300-μm sieve. This process removes fine particles, such as silt, that may fill the concavities of rounded insect sclerites, such as head capsules and the elytra of some weevils. If the silt is not removed, the affected sclerites may not rise to the top in the subsequent kerosene flotation procedure. Some fossil beetle workers prefer a slightly finer screen size (200–250 μm), and the study of smaller organisms requires even finer screens. Fossil midge larvae samples (Diptera: Chironomidae) are sieved on a succession of 200-μm and 100-μm screens (Hofmann, 1986), whereas fossil Cladocera (water fleas) are caught on 63-μm screens (Frey, 1986). If collaboration with researchers studying these animals is planned, either duplicate samples should be taken from the same horizon or the <300-μm fractions of samples should be retained for study by the other specialists.

Once the detritus has been screened, the residual material is placed, still damp, in a large bowl with a spout or a rectangular dishpan and processed by kerosene flotation to isolate and concentrate insect fossils. This procedure should always be performed in a room with good ventilation or under a fume hood. Kerosene or other lightweight oil is added to cover the sample and gently worked into the sample by hand for several minutes. The oil adheres to the smooth, impermeable insect sclerites but not to the more porous plant detritus. The remaining kerosene is decanted from the bowl and filtered into the stock bottle through a fine (200–300 μm) screen over a funnel. Since oil and water do not readily mix, the kerosene comes away cleanly from the mixture in the bowl and may be used repeatedly. Cold water is vigorously added to the oily detritus in the bowl, with the aid of a hose to reach the bottom of the sample and stir it thoroughly. In most samples, nearly all of the insect sclerites will rise to the top and float at the oil-water interface. Within 15–60 minutes, most plant residue sinks to the bottom of the bowl, and the now-concentrated insects may be decanted onto a 300-μm screen, to be washed gently in detergent and dehydrated in 95% ethanol before microscopic sorting.

In some peaty samples, notably those with abundant *Sphagnum* mosses, the plant macrofossils contain trapped air bubbles, which prevent them from sinking. Other plant fossils, such as the shiny seeds of some plants, float for the same reason as insect sclerites. Some success in sinking this noninsect material in water may be achieved by prolonged soaking of the samples in water for periods of a week or more. If this technique fails, or if both plant and insect fragments sink to the bottom in the flotation procedure, the entire sample must be sorted under the microscope. It is also necessary to check both sink and float fractions for fossils. Sometimes, for reasons not yet understood, the kerosene flotation fraction has a very low yield, even though the sample may be rich in insect fossils. One only learns this by checking both fractions. It should also be noted that the kerosene flotation method yields a sample rich in some insect body parts and poor in others. Not all insect parts are equally buoyant, even when coated with oil. For instance, the head capsules of heavily sclerotized beetles are sometimes too dense to float in water.

One variation on this procedure, used mostly with alluvial debris, lacustrine sediments, and peats, is to place the sample in water and, while waiting for it to become disaggregated, to decant all the material that floats in water through nested 5/20/80 mesh sieves. This step removes most large wood pieces and many of the plants that float in water (and that would also float in kerosene). It provides a concentrated sample of fossils, such as oribatid mites and chironomid heads, that would float in the kerosene flotation process but would be diluted by the abundance of other fossils.

Small samples should be sorted completely, rather than processed with kerosene. Samples that may need to be submitted for radiocarbon dating should also not be exposed to kerosene.

Kerosene flotation samples may be stored after washing. If a fungicidal detergent is used to wash the sample, it may be stored for long periods in water without suffering fungal growth, allowing the sample to be picked through in a dish of water, rather than in alcohol. However, picked specimens should be stored on a long-term basis in alcohol, since it retards fungal and bacterial attack.

Specimen sorting is carried out under a low-power (10×) binocular microscope in alcohol or water. Vials of specimens stored in alcohol tend to dry out unless impermeable stoppers or caps are used. Only ethyl alcohol (ethanol) should be used for sorting and storing insects. Methanol releases poisonous fumes, and isopropyl alcohol forms a cloudy solution when mixed with water. Some workers add a small amount of glycerine to the vials of ethanol to retard evaporation.

Robust beetle, bug, and ant fossils may be mounted with gum tragacanth (a water-soluble glue) on micropaleontology cards with cover slips and aluminum holders. However, many specimens shrivel and break during the drying of the glue on the card, so it is best to leave most if not all specimens in alcohol. Micropaleontology cards have a 3-mm-deep rectangular cavity, and the specimens are mounted on the floor of this cavity. This method works well for smaller specimens (up to a few millimeters in length), but large specimens,

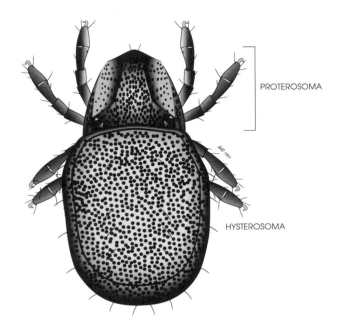

Figure 2.13. Generalized drawing of the dorsal surface of an oribatid mite.

such as those found in many packrat midden samples, will not fit under the cover slip and must be stored in vials of alcohol, or on cards with deeper cavities.

Fossil caddisfly larval sclerites are examined with transmitted light. Oribatid mite and chironomid fossils are also usually cleared and mounted on glass slides for viewing with a transmitted light microscope. Methods used in the study of caddisfly fossils are summarized by Williams (1988).

Many exoskeletons of oribatid mites (Fig. 2.13) are found nearly intact in Quaternary peats and lake sediments, especially from sediments deposited under cold climatic regimes. The extraction and preparation of oribatid mites for identification involves a different set of laboratory procedures, as discussed by Erickson (1988). However, standard 300-μm mesh sieving and kerosene flotation often yields substantial numbers of oribatids.

Special procedures are required to extract and identify fossil chironomid larvae remains from lake sediments. The principal body part preserved is the head capsule, and these head capsules may be very abundant in lake sediments. Walker (1987) reported that sediments from lakes in forested watersheds generally yield 50–100 fossil head capsules per milliliter. However, inorganic sediments, such as those frequently encountered in late glacial deposits, yield far fewer specimens. Fortunately, the head capsules of midges contain many diagnostic features.

Samples are usually deflocculated in 5–10% KOH prior to analysis. Head capsules are then separated from finer debris by sieving through screens of 100-μm mesh or finer. The residue is then stored in alcohol and sorted in alcohol under a binocular microscope, using 50× power (roughly five to ten times the magnification used when sorting fossil beetle remains).

3

IMPORTANT FOSSIL INSECT GROUPS AND THEIR IDENTIFICATION

> The most useful characters of subfossil remains of Coleoptera, especially
> when consisting merely of single or fragmentary elytra, lie in the micro-
> sculpture.
>   —Carl H. Lindroth (1948)

The identification of most insect fossil sclerites from Quaternary deposits is a
painstaking task, made more difficult by a general lack of suitable identification keys. Most dichotomous keys written for the identification of modern
insects require entire specimens, or even a series of specimens of both sexes.
Comparable fossil material is rarely available.

Fortunately, there are some exceptions to this rule. Some monographs on
beetle families and genera include numerous illustrations, including useful
photographs and line drawings, either of complete specimens or of prominent
sclerites (head capsules, pronota, and elytra) that are also found in the fossil
record. Carl Lindroth's (1961, 1963, 1966, 1968, 1969) series on the ground
beetles of Canada and Alaska and his revision (1985, 1986) of the Fennoscan-
dian ground beetles are good examples of systematic publications that are very
useful to paleoentomologists. Other beetle taxonomists who have provided
valuable illustrations in their generic revisions include Milton Campbell, Henri
Goulet, and Aleš Smetana at the Biosystematics Research Centre, Agriculture
Canada, Ottawa, who have included descriptions and scanning electron micrographs (SEMs) of major sclerites frequently found as Quaternary fossils.
Smetana's (1985) revision of North American *Helophorus* (Hydrophilidae)
species includes SEM plates and detailed descriptions of the pronotum of each
species. Goulet's (1983) revision of Holarctic Elaphrini (Carabidae) includes
many SEM photos of head capsules, pronota, and elytra. Likewise, Campbell's
revisions of the North American rove beetle genera *Acidota* and *Olophrum*
(Campbell, 1982, 1983) include SEM photos and taxonomic discussions of the
features observable on head capsules, pronota, and (in the case of *Acidota*)
elytra.

It is also possible to gain some familiarity with regional faunas by looking through monographs with abundant, accurate illustrations of them. Browsing through a series of good photographs or line drawings may provide clues useful in the identification of an unknown specimen. Matching a fossil specimen to an illustration (either exactly or nearly so) can be an important first step in its identification, narrowing the field of possibilities and thus greatly reducing the amount of time then needed to compare fossil and modern specimens in a museum collection.

Most Quaternary insects are identified through direct comparison with modern, identified material. It is perhaps no coincidence that the first successful attempts at identifying Quaternary insect species took place in Britain. The British fauna is relatively small (comprising only a few thousand species), and a synoptic collection of British beetles can be placed on a tabletop. Coope and Osborne have such a collection in their laboratory in Birmingham. Of course, having a collection of beetles is very different from having a good knowledge of the species in that collection. It is necessary to develop familiarity with the insect fauna that currently lives in a study region before attempting to identify insect fossils from that region. Furthermore, the piecing together of a paleoecological scenario based on insect assemblages from a study region must be based on a sound knowledge of the ecological requirements and interactions of the species found in the fossil assemblage. Although some information may be gleaned from the literature, there is no substitute for prolonged study of modern material and hands-on experience, gained by years of observing, collecting, and identifying modern beetles from a given region.

If Coope had started working elsewhere, he might not have achieved such early success. The beetle fauna of North America comprises more than 30,000 species in 98 families (White, 1983). The fauna of Eurasia is at least as large. It is difficult to comprehend the extraordinary diversity of insects, except perhaps by comparison with other groups of organisms. In contrast, Anderson (1984) has estimated that all the mammals of the Quaternary (both extinct and extant) comprise about 500 species. A mammologist might be competent to speak about the mammalian fauna of an entire continent. On the other hand, no single entomologist can be expected to retain a comprehensive knowledge of the beetle fauna of a continent. Therefore paleoentomologists seek the assistance of specialists in various families. Most of these taxonomists are housed at national museums of natural history. Other specialists may be scattered in colleges and universities or in government agricultural and forestry offices. The budding Quaternary entomologist must develop a network of contacts in order to find out who knows how to identify which groups and who is willing to look at fossil material. In the end, of course, it remains the worker's responsibility to verify any identifications made by taxonomic specialists.

The fauna of any single region is necessarily much smaller than the fauna of a whole continent, especially if the study region is in the higher latitudes, where species diversity is greatly diminished. I have found that, after a few years of working on the fossil fauna of a given region, it is possible to develop enough familiarity with the taxa to be able to identify most of the fossil material without the help of specialists working on the modern fauna.

Fortunately, the exoskeletons of beetles, ants, caddisfly larvae, and some other insects and arachnids exhibit a multitude of features useful in the assignment of fossils to orders, families, genera, and species. The general success rate for the identification of fossils of major body parts (head capsules, thoraxes, and elytra) to the species level is about 50%. That is, about half of the major sclerites in a given fossil assemblage will eventually be confidently identified to species. This success rate varies from region to region. British workers have had far greater success with some samples. For instance, Coope and Angus (1975) identified 252 species, or nearly 90%, of a rich beetle fauna comprising 282 taxa from the Isleworth site in Britain. Of course, many insect body parts, such as leg segments, antennal segments, and abdominal sclerites, cannot be identified even to the family level. The investigator quickly learns which types of specimens are worth studying in detail and which are not.

USEFUL CHARACTERS FOR FOSSIL BEETLE IDENTIFICATION

One reason that beetles preserve so well in sediments is that many species' exoskeletons form a hard, armorlike case over their bodies. The major sclerites, especially on the dorsal side, are not just flimsy body coverings, but large, heavily sclerotized plates (Fig. 3.1). The main exoskeletal parts that preserve in sediments are the head capsule, the pronotum (the dorsal thoracic shield; Fig. 3.2), and the wing covers, or elytra (Fig. 3.3). Many of these plates are ornamented with a number of complex features that are preserved for as long as the sclerites themselves are preserved. These characters include striae (either isolated or in rows), carinae (Fig. 3.4B), grooves, tubercles, punctures, setae (or points of attachment of setae, if these have broken off), scales (Fig. 3.4A), and rows of teeth. Each of these features represents character states that may exhibit a wide range of variation from one taxon to another. For instance,

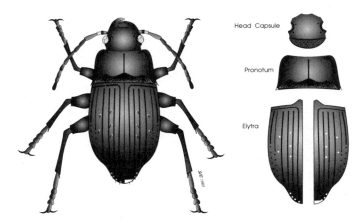

Figure 3.1. Generalized drawing of the dorsal surface of a ground beetle (Carabidae), showing sclerites frequently preserved as Quaternary fossils.

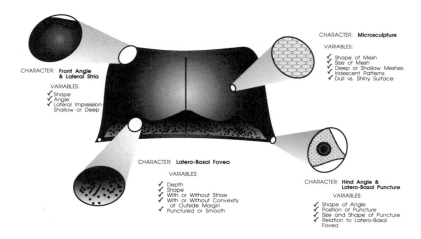

Figure 3.2. Generalized drawing of a ground beetle pronotum, showing a range of diagnostic features used in fossil identification.

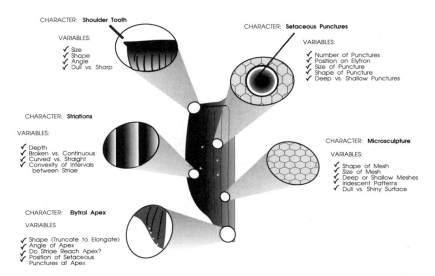

CHARACTER: **Shoulder Tooth**

VARIABLES:
✓ Size
✓ Shape
✓ Angle
✓ Dull vs. Sharp

CHARACTER: **Setaceous Punctures**

VARIABLES:
✓ Number of Punctures
✓ Position on Elytron
✓ Size of Puncture
✓ Shape of Puncture
✓ Deep vs. Shallow Punctures

CHARACTER: **Striations**

VARIABLES:
✓ Depth
✓ Broken vs. Continuous
✓ Curved vs. Straight
✓ Convexity of Intervals
  between Striae

CHARACTER: **Microsculpture**

VARIABLES:
✓ Shape of Mesh
✓ Size of Mesh
✓ Deep or Shallow Meshes
✓ Iridescent Patterns
✓ Dull vs. Shiny Surface

CHARACTER: **Elytral Apex**

VARIABLES
✓ Shape (Truncate to Elongate)
✓ Angle of Apex
✓ Do Striae Reach Apex?
✓ Position of Setaceous
  Punctures at Apex

Figure 3.3. Generalized drawing of a ground beetle elytron, showing a range of diagnostic features used in fossil identification.

punctures may range from deep to shallow and from broad to narrow; they may contain a seta or they may not; they may occur in dense patches, in rows, or in small groups, or may be widely dispersed; and they may vary in shape from round to oblong to quadrate.

Microsculpture, some of which is visible only at high magnification (150–200×), is part of the exoskeleton of most insects. As Lindroth (1948) observed, microsculpture is very useful in identifying fossil beetles, because the microlines do not degrade or alter with time, and are remarkably constant within species. Most microsculptural details are best viewed using a light source aimed from a very oblique angle. Unfortunately, bright light reflecting on the shiny surface of a specimen makes viewing of microsculpture difficult. This problem can be overcome by setting a small screen of opaque Mylar film between the light source and the specimen, as described by Campbell (1979). The screen effectively diffuses the light.

A few beetles are devoid of microsculpture, giving them a shiny appearance. Others have patches of microsculpture only at the margins of the pronotum and the apexes of the elytra. Many species are completely covered with very dense patterns of reticulate lines, often in diagnostic shapes. These range from isodiametric meshes (Fig. 3.5A) to longitudinally or transversely elongated or broken meshes (Fig. 3.5B–D). In some species, the meshes are broken into transverse rows of very fine lines (Fig. 3.5E). These serve as diffraction

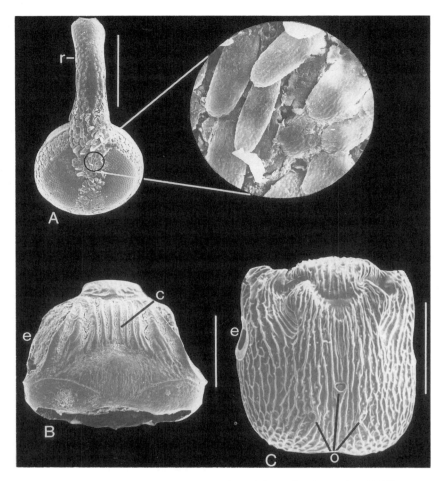

Figure 3.4. Scanning electron micrographs of fossil insect head capsules. (A) The weevil *Cylindrocopturus armatus*, showing rostrum (r) and preservation of scales (inset). (B) The ground beetle *Notiophilis borealis*, with longitudinal carinae (c) on the frons and eye socket (e). (C) The ant *Myrmica incompleta*, showing eye socket (e) and primitive eye spots or ocelli (o). Scale bars equal 0.5 mm. (Photographs by the author.)

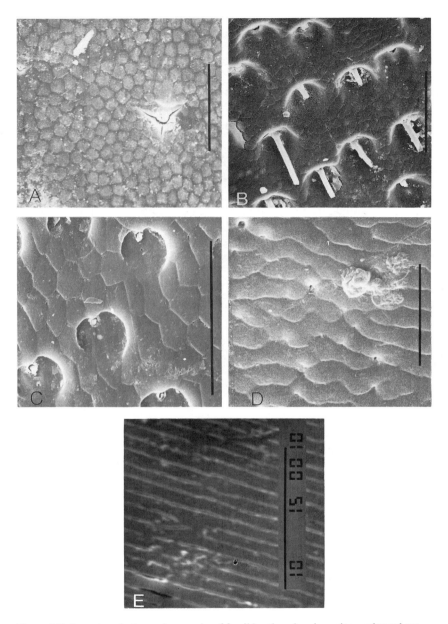

Figure 3.5. Scanning electron micrographs of fossil beetles, showing microsculptural patterns grading from isodiametric mesh (A) through elongated meshes (B and C) to broken meshes (D) and transverse lines (E). (A) Microsculpture on elytron of *Onthophagus* sp. (Scarabaeidae). (B) Microsculpture on pronotum of *Thanatophilus truncatus* (Silphidae). (C) Microsculpture on elytron of aleocharine staphylinid. (D) Microsculpture on pronotum of *Tachinus brevipennis* (Staphylinidae). (E) Microsculpture on elytron of *Selenophorus gagatinus* (Carabidae). Scale bar equals 50 μm on A–D, 10 μm on E. (Photographs by the author.)

gratings that cause iridescence. Other beetles have a very dull, mottled appearance because of dense, granular microsculpture and ornamentation.

Microsculpture is thought to serve a variety of functions in insects, as discussed in Crowson (1981, pp. 300–302). Iridescence is thought by some authors to provide the beetle with some measure of protection from solar radiation, since it is most common in species living in low latitudes (Lindroth, 1974). Some microsculpture serves a cryptic function. One species of ground beetle, *Agonum bembidioides,* has alternating patches of short microsculptural lines at oblique angles to each other, producing alternating patterns of silver and black spots. As Lindroth (1966) noted, this peculiar attire serves as extremely efficient camouflage for the beetle, greatly resembling as it does the opalescent pieces of charcoal in the newly burned forests that it inhabits.

Metallic coloration is fairly common in beetles, especially in some families (e.g., Cicindelidae, Carabidae, Scarabaeidae, Buprestidae, Meloidae, and Chrysomelidae). These metallic colors are called structural colors, because they result from light interference with the thin films of the cuticle. These colors persist even when the outer waxy layers of epicuticle are removed; they are probably derived from the outer part of the endocuticle (Crowson, 1981, p. 303). Some metallic-colored fossil specimens become darkened through time, but the metallic sheen reappears when the fossils are wetted. In such specimens, the colors may shift in hue. For instance, if a beetle was originally a metallic green its fossil sclerites may show a blue-green color when wetted (Coope, 1959). Bright coloration probably serves as a defense mechanism in insects. If a metallic or iridescent beetle runs from the shade into bright sunlight, it flashes a bright, startling color that may surprise a predator, thus allowing the beetle to elude it.

Another form of coloration is due to pigments. Pigment-based colors may or may not be preserved in insect fossils. Because of the instability of some pigments, sclerites that were originally yellow, orange, and red tend to thin and become frail through time, whereas dark brown or black coloration owing to tanning of chitin may persist (Coope, 1959). Hence, the black spots on a ladybird beetle (Coccinellidae) may preserve longer than the red or orange background, which disappears, leaving only the spots, like the smile on the Cheshire Cat. However, pigmented ornamentation such as spots, stripes, and maculae persists well enough in most fossils to serve as an aid to identification. In fact, coloration patterns are sometimes diagnostic in identifying species of ground beetles, predaceous diving beetles (Dytiscidae), ladybird beetles (Coccinellidae), leaf beetles (Chrysomelidae), and others (Fig. 3.6).

Figure 3.6. Light microscope photographs of fossil beetles, showing preservation of pigmented color patterns. (A) pronotum of *Hippodamia convergens* (Coccinellidae). (B) elytron of *Hyperaspis trifurcata* (Coccinellidae). (C) elytron of *Zygogramma tortuosa* (Chrysomelidae). Scale bars equal 0.5 mm. (Photographs by the author.)

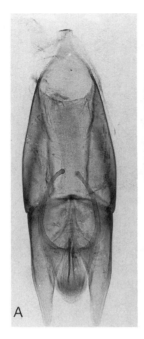 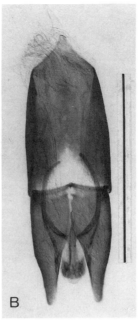

Figure 3.7. Light micro-scope photographs of fos-sil (A) and modern (B) aedeagi of *Helophorus aquaticus.* The fossil specimen is from the Starunia site in the Ukraine (see Chapter 1). It was originally identified as *Helophorus dzieduszckii* by Lomnicki (1894). Both × 100. Scale bar equals 0.5 mm for both. (Photographs courtesy Robert Angus.)

Generally, the ventral abdominal sclerites become separated from each other and isolated in sediments. However, whole abdomens of beetles are occasion-ally preserved intact. Inside the abdominal exoskeleton of male beetles is the aedeagus, or male genitalia (Fig. 3.7). This is also partially sclerotized, and so preserves well in many organic sediments. In addition, the terminal abdominal sclerites in some rove beetles, such as those of the genus *Tachinus,* are pro-duced into a series of elongated lobes (Fig. 3.8). The shapes of these sclerites are the principal diagnostic features used to identify both modern and fossil specimens.

The head capsules of many families of beetles contain diagnostically useful features, including the shape, position, and size of the eye sockets; the position and shape of the mouth parts (many mandibles are found separately from head capsules, but their position and points of attachment on the head can be discerned even when they are absent); the shape of the frontal region (that between the eyes); and the shape of the clypeus (the part of the head anterior to the frons or frontal sclerite). In ground beetles, the number and position of setaceous punctures inside the margin of the eyes is important, and some ground beetles have longitudinal ridges or carinae extending from the base of the frons to the clypeus (Fig. 3.4B). The length, shape, and number of these ridges is a diagnostic feature for some genera. The number and position of

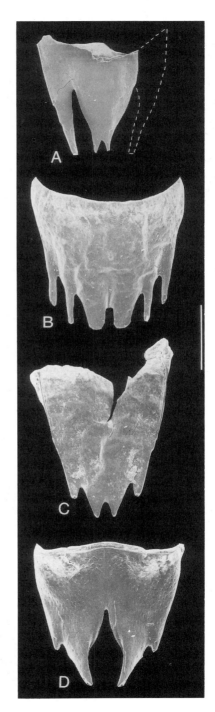

Figure 3.8. Scanning electron micrographs of terminal abdominal segments of the rove beetle *Tachinus nearcticus* (fossil specimens). (A) Eighth female abdominal tergite. (B) Eighth female abdominal sternite. (C) Eighth male abdominal tergite. (D) Eighth male abdominal sternite. Scale bar equals 0.5 mm. (Photographs by the author.)

setaceous punctures (chaetotaxy) is also an important character in the identification of some rove beetles, such as the genus *Tachyporus* (Campbell,
1979). In weevils (Curculionidae), the front of the head is produced into a snout
or rostrum. The elongated first segments of weevil antennae generally fit into
grooves along the snout. The size and shape of the snout and the shape and
position of antennal grooves are important taxonomic features.

The dorsal prothoracic shield or pronotum is often useful in beetle identification. In some families, such as most ground beetles and predaceous diving
beetles (Dytiscidae), the pronotum is often a broad, flat plate that is completely
divided from the ventral part of the prothorax. This means that fossil pronota of
these families are usually found detached from their associated thoracic sclerites. In other families, including most weevils, the pronotum is cylindrical and
fixed to the ventral sclerites, forming a domelike covering over the thorax.
Some beetle pronota have straight lateral margins; others have sinuate margins.
Some pronota are smooth and shiny; others are heavily sculptured and rugose
or covered with scales or setae. The overall size, shape, and texture of many
pronota provide enough information to enable an identification to the family or
genus level. Pronota of ground beetles offer the best characters for species
determinations; only the male genitalia are more diagnostic.

Elytra are the modified forewings of beetles that serve as reinforced, protective covers for the hind wings and often the abdomen. In fact, the name of the
beetle order Coleoptera comes from the Greek term for "sheath-winged."
Elytra are the single most important feature that characterizes beetles as a
group, and they have undoubtedly contributed to the success of beetles in terms
of abundance and diversity. It is advantageous for an insect to have a protective
covering over the top of an otherwise somewhat fragile abdomen with its vital
organs. Elytra may be pliable and leathery (as in the Cantharidae, Lampyridae,
and Meloidae) or very hard and inflexible (as in many weevils). Few leathery
elytra are preserved as Quaternary fossils. As with pronota, beetle elytra range
from relatively flat to very convex. Most beetle families have elytra that taper
apically and cover the abdomen (or at least all but the last one or two segments). However, some families (Staphylinidae, Hydroscaphidae, some Silphidae, and Histeridae) have truncated elytra that leave several abdominal
segments exposed. The usefulness of elytral characters for fossil identification
varies greatly from family to family and from genus to genus. Some elytra
contain sufficient diagnostic characters to allow a species identification; others
may not even be identifiable with any certainty to the genus level.

Different depositional environments (e.g., fluvial versus lacustrine deposits)
affect the numbers and types of insect body parts in assemblages. For instance,
the head capsules and pronota of weevils are rarely deposited in the same

locality as their elytra, because of differences in buoyancy (Morlan and Matthews, 1983). Likewise, buoyancy differences cause some insect fragments to be under-represented in kerosene flotation. Sediment retained in head capsules or thoraxes causes them not to float as well as do body parts that are free of sediment.

IDENTIFICATION OF OTHER GROUPS OF FOSSIL INSECTS

Several other insect orders are frequently encountered in Quaternary sediments. Many ants (Formicidae) have heavily sclerotized head capsules that contain sufficient diagnostic characters to allow specific identifications (Fig. 3.4C). Ant mandibles are heavily sclerotized, and are often found in fossil assemblages. They are not frequently identified to species level, but even genus identifications may be of use in paleoenvironmental reconstructions. For instance, the mandible of a carpenter ant (genus *Camponotus*) offers evidence of the availability of wood at the fossil site at the time of deposition, since carpenter ants make their nests in rotting logs. Identification of ant fossils is complicated by the fact that ants are grouped into castes (queens, males, and workers), the head capsules of which are different. Most ants are workers, but the head capsules of queens have been found in some assemblages (e.g., Francoeur and Elias, 1985; Mackay and Elias, 1992). Ants, like many beetles, are general predators and scavengers not tied to specific prey or host plants; their fossils supply valuable information on past climatic conditions. For instance, in northern studies, ants provide evidence of conditions suitable for the growth of trees, since no ants are known to inhabit regions well beyond the arctic treeline (Francoeur, 1983; Gregg, 1972). Their fossils may, however, occur in samples representing subarctic conditions.

Head capsules, thoraxes, and the sclerotized portion of the hemelytra of bugs (Hemiptera and Homoptera) are also common in Quaternary sediments. Fossils of stink bugs (Pentatomidae), seed bugs (Lygaeidae), and leaf hoppers (Cicadelidae) are fairly common in terrestrial sediments, and water striders (Gerridae), shore bugs (Saldidae), water boatmen (Corixidae), and backswimmers (Notonectidae) are found in fluvial and lacustrine sediments. Recently, packrat midden assemblages from the Chihuahuan Desert have yielded head capsules and thoraxes of Cydnidae (burrower bugs; Fig. 3.9), and also head capsules of some Reduviidae (assassin bugs) that parasitize packrats (Elias and Van Devender, 1990, 1992).

The larvae of caddisflies (Trichoptera) are aquatic, and sclerites from the head capsule and thorax of caddisfly larvae are abundant in some lacustrine sediments (Fig. 3.10). The frons and clypeus of caddisflies are fused into a

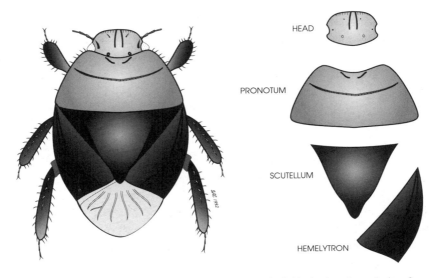

Figure 3.9. Generalized drawing of a burrower bug (Cydnidae), showing sclerites frequently preserved as Quaternary fossils. (After Froeschner, 1960.)

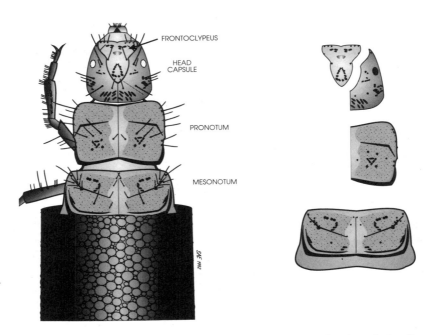

Figure 3.10. Sketch of the head and thorax of a caddisfly larva, showing sclerites frequently preserved as Quaternary fossils. (After Wiggins, 1977.)

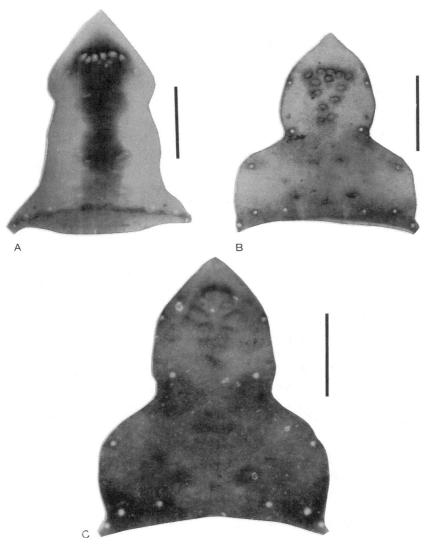

Figure 3.11. Light microscope photographs of fossil caddisfly larval fossils.
(A) Frontoclypeus of *Agrypnia crassicornis* (Phryganeidae). (B) Frontoclypeus of
Asynarchus lapponicus (Limnephilidae). (C) Frontoclypeus of *Potamophylax* sp.
(Limnephilidae). Scale bars equal 0.5 mm. (Photographs by the author.)

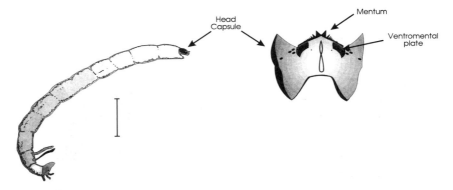

Figure 3.12. Generalized drawing of a midge (Chironomidae) larva, showing taxonomically useful features of the head capsule. Scale bar equals 1 cm. (After Borror and White, 1970.)

single sclerite, called the frontoclypeus. Besides its overall shape, the frontoclypeus possesses a number of other diagnostic features, including the position of setaceous punctures around its margins, surface sculpturing, and the size, shape, and coloration of muscle scars that are revealed under transmitted light (Fig. 3.11). The pronotum and mesonotum are also useful for identification (Williams, 1988). The cases built by caddisfly larvae are occasionally found in ancient lake sediments. The size and shape of the cases, as well as the materials used in their construction, are often diagnostic to the family or genus level. Though more rare, fossils of adult caddisflies have also been found.

Caddisfly larvae provide valuable information on the waters they inhabit, as many species have narrow thermal tolerances and are sensitive to the trophic status and pH of the water (Wiggins, 1977). Some species require specific substrates and build larval cases from particular substances (e.g., sand grains of a certain size, reeds, and even snail shells).

The study of fossil midge larvae (Diptera: Chironomidae) also plays an important part in paleolimnology. The larvae are aquatic, and, as with caddisfly larvae, the species composition of midges in lake sediments can provide a great deal of information on water quality and substrates (Walker et al., 1991b). Most diagnostic features are preserved on the head capsule (Fig. 3.12). Phantom midge larvae (Diptera: Chaoboridae) have also been used as paleolimnological indicators (Uutala, 1990).

Fossil midge samples should be taken from the center of a lake, rather than from the margins, where fossil beetle samples are taken (Hofmann, 1986). Midge head capsules occur in such great numbers in many lake sediments that even a small-diameter piston core of the type often used by palynologists

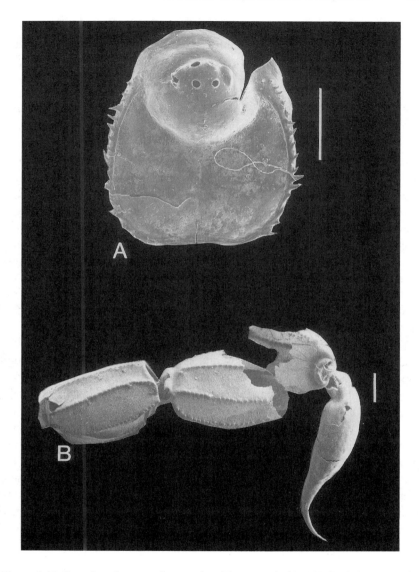

Figure 3.13. Scanning electron micrographs of fossil arachnids. (A) Cephalothorax of the spider *Erigone dentiger*. (B) Tail segments of a scorpion. Scale bars equal 0.5 mm. (Photographs by the author.)

provides samples with sufficient numbers of specimens for a detailed analysis. This is why the study of chironomid fossils is often performed in concert with pollen analysis, as in the study of paleoenvironments at Marion Lake, British Columbia, by Walker and Mathewes (1987).

Mites (Arachnida: Acarina), especially oribatid mites, though technically not insects, can be treated in the same way and are an important element of the soil fauna of many regions, especially at high latitudes. Many oribatid mites are preserved more or less intact in sediments, because they are generally small (<0.5 mm) and compact, with short appendages. Recently, the study of fossil oribatid mites has begun to blossom, as scientists seek to understand ancient substrates and paleosols (Erickson, 1988; Krivolutsky and Druk, 1986; Krivolutsky et al., 1990). Mites are an important part of the soil fauna of arctic and alpine ecosystems, and their remains are often very abundant in sediments from those regions. Like chironomids, oribatid mites are so abundant in some sediments that they can be obtained from samples collected from small-diameter piston cores.

The more heavily sclerotized parts of arachnids are occasionally preserved in Quaternary deposits. Spider (Araneida) cephalothoraxes are not uncommon (Fig. 3.13A). Leech and Matthews (1971) described a fossil crab spider from a late Tertiary deposit in Alaska. Some packrat middens contain claws and tail segments of scorpions and pseudoscorpions (Fig. 3.13B). The genital structures of spiders are also occasionally found. The male palp offers characters for specific identification that are probably as valuable as those found in the male genitalia of beetles and other insects.

The exoskeletal remains of insects and other arthropods are abundant and well preserved in many Quaternary sediments. Their identification is facilitated by the variety of diagnostic features preserved on the fossil sclerites. The major obstacle in most fossil insect identifications is the investigator's lack of familiarity with the taxon in question. Nevertheless, as John Matthews (written communication, 1992) has observed, there is no mystique to the identification of insect fossils. All it takes is patience and a good eye for detail.

4

THE VALUE OF INSECTS IN PALEOECOLOGY

> In number of described species beetles represent the largest group of organisms . . . their role in the operation of ecosystems, particularly on land, should never be underestimated.
>
> —M. Ghilarov, in Crowson (1981)

ABUNDANCE AND DIVERSITY OF INSECT ASSEMBLAGES

Insects are arguably the most successful group of organisms ever to have inhabited the earth. The present-day crop of insect orders has been in existence since long before the age of dinosaurs, and they represent roughly three-quarters of all animal species known at present. Insect abundance and diversity are important elements in their fossil record, because the fossil assemblages are likewise often abundant and diverse, providing an unusually rich record of past life.

This chapter focuses on beetles because they are the most important insect order in the Quaternary fossil record. They are also the most diverse order, with more than 300,000 known species (White, 1983); about 1500 new species are described each year (Arnett, 1973). Beetles account for 25% of all known species of organisms, a quantity that is more than all flowering plants combined. Little wonder then that when the great defender of evolutionary theory, J. B. S. Haldane, was asked what the observation of nature might reveal of the mind of the Creator, he responded, "an inordinate fondness for the Coleoptera" (Fisher, 1988, p. 313).

The diversity of beetles is difficult to comprehend. Crowson (1981) lists 168 families of beetles. This taxonomic variety corresponds to an equally diverse ecological complexity (Fig. 4.1). Beetles occupy almost every conceivable ecological niche and type of habitat on land and in fresh water. This diversity makes them an important group in the fossil record, because their remains serve as proxy data for a wide variety of habitats and environmental conditions.

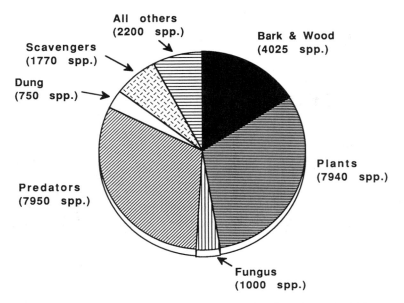

Figure 4.1. Percent composition of North American beetle fauna by ecological group.
Bark and wood feeders are mostly in the families Scolytidae, Curculionidae, Bupres-
tidae, and Cerambycidae; plant feeders occur in many families, but especially
Chrysomelidae and Curculionidae; fungus feeders are part of the Mycetophagidae,
Ciidae, Erotylidae, and others; predators include members of the Cicindelidae,
Carabidae, Dytiscidae, and Staphylinidae, as well as some other families; dung
feeders are mostly in the family Scarabaeidae; scavengers include members of the
Dermestidae and several other families. (Data derived from White, 1983.)

They are found from arctic polar deserts (Danks, 1981) to the subantarctic
islands (Crowson, 1981), and at elevations as high as 5600 m in the Himalayas
(Mani, 1968).

Beetles live in a bewildering variety of specialized habitats; thus their
fossil remains provide a richness of paleoecological detail. Dead trees and
associated fungi nurture hundreds of different species of beetles. Fossils of
these species often survive to tell of ancient old-growth forests, even if the
trees themselves have not been preserved. Besides living in lakes, ponds,
and streams of all sizes, some aquatic beetles also inhabit hot springs and
brackish water.

Some ground beetles and rove beetles live in mammal and bird nests, where
they prey on fleas, mites, and other parasites. The larvae of carrion beetles
(Silphidae) feed on carrion, but adults also prey on other insects (especially
maggots) found on carrion (Anderson and Peck, 1985). Again, the fossil

remains of these insects confirm that their avian and mammalian hosts themselves were present at the time and place in question, whether or not the hosts themselves were preserved as fossils.

Predatory beetles are an important component in most fossil assemblages, and they have much to tell us about past environments. They are found in several families, including tiger beetles (Cicindelidae), ground beetles (Carabidae), predaceous diving beetles (Dytiscidae), rove beetles (Staphylinidae), hister beetles (Histeridae), ladybird beetles (Coccinellidae), and checkered beetles (Cleridae). Most ground beetles prey on a wide variety of arthropods and other invertebrates, but some genera specialize in one type of prey. Beetles of the genus *Calosoma* are caterpillar hunters, as are some species of *Calleida*. *Cychrus* feeds on snails. *Dyschirius* is a riparian genus that burrows in soil, where it preys on a soil-burrowing rove beetle, *Bledius,* and the variegated mud-loving beetles (Heteroceridae) that burrow in moist stream banks. The genus *Lebia* climbs plants to prey on other insects. Some species prey exclusively on certain leaf beetles (Chrysomelidae), and some *Lebia* species mimic the appearance of their prey species (Lindroth, 1969). Ladybird beetles feed on plant lice and scale insects so effectively that they are used as biological control agents for these plant pests. Many checkered beetles live under bark, where they attack bark beetles (Scolytidae).

Although predaceous and scavenging beetles show considerable diversity and degree of specialization, the abundance and diversity of phytophagous beetles are even greater. Plant-feeding beetles are an important element in fossil reconstructions, because they provide information on past plant communities, including such aspects as the composition of plant communities (both aquatic and terrestrial) and the health and age-class structure of tree stands. The majority of phytophagous beetles are found on the leaves and flowers of plants (7940 North American species in 25 families), and a considerable number live under the bark of trees or in rotting wood (4025 North American species in 24 families). Some phytophages are generalists, but many are associated with a few or only one host plant species. The fossils of these beetles provide data on ancient plant communities, even when the plants' pollen signature is sparse or lacking (Elias, 1982a).

The numerous small families of beetles serve to round out the paleoenvironmental scenario derived from fossil assemblages. There are families that specialize in feeding on mosses, fungi, slime mold, sap, algae, and pollen (a total of about 1300 North American species in 20 families). Dung beetles (a part of the family Scarabaeidae) can also either be generalists or feed on the dung of only a few species or just one species of vertebrate. The presence of dung beetles in a fossil assemblage may show that the animal that produced the dung was also a part of the ancient community, whether or not any fossil remains of the dung-maker are actually preserved.

Soil-dwelling insects in fossil records reveal types of past substrates and degree of soil development. Many types of beetles spend at least one part of their life cycle in the soil. Some beetle groups are adapted to particular soil types. For instance, many tiger beetles (Cicindelidae) are found only on sandy soils. Ground beetles that prey exclusively on snails frequent calcareous soils. Some beetles burrow deeply into soil cracks; others are adapted for a completely subterranean existence in caves. Many of the latter are blind, lacking in pigment, and dorsoventrally compressed (a character that allows them to penetrate narrow cracks more easily).

As we have seen, insect fossil records are not limited strictly to water-lain sediments in the middle to high latitudes. Packrat middens and cave deposits in desert regions are now yielding important information based on fossil insect assemblages (Elias, 1990a). The dried dung of Pleistocene mammals has been preserved in many caves in arid regions as well. Waage (1976) reported on fly larval and pupal remains (family Sciaridae) from dried dung in Gypsum Cave, Nevada. The dung was apparently that of the Shasta ground sloth, *Nothrotheriops shastensis*.

Desert-dwelling beetles show some remarkable adaptations to life in hot, dry environments. Darkling beetles (Tenebrionidae) have done particularly well in the world's deserts. The exceptionally thick, impermeable exoskeleton of some darkling beetles enables them to reduce moisture loss. One North American species covers its body with a thick secretion of wax during times of extreme aridity; the wax exhibits distinct color phases, ranging from light blue under conditions of low humidity to jet black under high humidity. The wax coating cuts water loss from the cuticle, and the pale blue color exhibited during arid conditions lowers the body temperature by decreasing absorption of solar energy (Hadley, 1979).

The arctic and alpine regions serve as an excellent repository for insect fossils. The adaptations of cold-hardy beetles are just as impressive as those of their desert-dwelling counterparts. Cold hardiness is made possible by a variety of metabolic responses to the changing seasons, as discussed in Danks (1978). The liquids in some beetles are able to supercool to temperatures as low as –40°C, preventing ice formation through the masking or removal of particles that could serve as nuclei for ice crystals. Other beetles are freeze-tolerant, that is, they are able to survive extracellular freezing within their bodies. These animals secrete nucleators (proteins and polypeptides) that actually promote the growth of ice crystals at temperatures above 0°C. This process in turn prevents the rapid formation of ice crystals in very cold supercooled liquids.

Such ice formation may rupture cells, whereas the controlled freezing of extracellular liquids at temperatures above the normal freezing point of water prevents this kind of damage.

SPECIES CONSTANCY IN THE QUATERNARY

Ernst Mayr (1970), commenting on the limits of our knowledge of speciation, observed

There is perhaps no other aspect of speciation about which we know as little as rate. Indeed, we shall probably never have very accurate information on this phenomenon. The splitting of one species into two is a short-time event that, as such, is not preserved in the fossil record. For information we rely entirely on inference.

The fossil record that Mayr considered when he made this statement was essentially the pre-Quaternary fossil record: laid down in bedrock, full of temporal and spatial gaps, and generally lacking in continuity. However, the Quaternary insect fossil record, albeit far from perfect, offers abundance of data, diversity of species, and chronological control (by radiocarbon dating and other radiometric methods) for many families, especially beetles.

As discussed in Chapter 1, the early Quaternary entomologists viewed all fossil beetles as extinct species, assuming that rapid insect speciation had accompanied the extreme climatic fluctuations and the waxing and waning of the continental ice sheets of the Pleistocene. In fact, the fossil beetle record of the middle and late Pleistocene provides no evidence of speciation (Coope, 1970; Matthews, 1976a,b). Unlike the case of the Pleistocene mammalian megafauna, there is no evidence for any significant beetle extinction events in the Quaternary.

Some of the best evidence for beetle species longevity comes from late Tertiary and early Quaternary assemblages in Alaska and the Canadian arctic, as summarized in Matthews (1977a, 1979a,b, 1980a). Matthews (1970) described two extinct species of the rove beetle genus *Micropeplus* from Pliocene deposits at the Lava Camp Mine, Alaska. He believed *M. hoogendorni* (Fig. 4.2) and *M. hopkinsi* to be the immediate precursors of the modern species. These fossils are about 5.7 million years old. However, Holderidge (in Coope, 1987b) found specimens of *M. hoogendorni* in British fossil assemblages from the middle Pleistocene, and the Russian entomologist Rjabukhin has described a Siberian species, *M. dokuchaevi,* that appears to be synonymous with *M. hoogendorni* (J. M. Campbell, personal communication, 1992).

Figure 4.2. Scanning electron micrograph of fossil elytron of the rove beetle, *Micropeplus hoogendorni*. (From Matthews, 1976b. Courtesy Geological Survey of Canada, Department of Energy, Mines and Resources. Reproduced with the permission of the Minister of Supply and Services Canada.)

Matthews also described two species of the water scavenger beetle (Hydrophilidae) genus *Helophorus* from the Lava Camp Mine and another Pliocene-age deposit (Beaufort Formation) on Meighen Island in the western part of the Canadian arctic archipelago. The extinct species are the precursors of the extant species *H. tuberculatus,* a species found today in the arctic. It should be noted that Matthews also found numerous beetle fossils in these Pliocene deposits that match modern species.

Matthews (1977b) postulated a gradual reduction in wing size in the flightless rove beetle, *Tachinus apterus,* using a chronological sequence of Alaskan fossils spanning the Quaternary.

Additional late Tertiary beetle fossils have been described by Böcher (1989a, 1990) from Kap København, Greenland. The fossils probably date to the

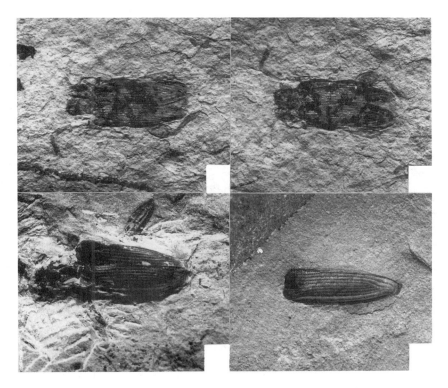

Figure 4.3. Fossil specimens of the leaf beetle, *Plateumaris nitida,* reidentified by Askevold (1990) from the Florissant shales in Colorado. (Courtesy National Research Council of Canada.)

Pliocene-Pleistocene transition. They are remarkable not only for their excellent state of preservation, but also because they represent a boreal forest environment in northernmost Greenland. Nearly all of the insect species (beetles and ants) are extant.

Perhaps the most startling evidence for species longevity comes from the fossil record of the aquatic leaf beetles in the subfamily Donaciinae (Chrysomelidae). Askevold (1990) analyzed donaciine fossils from the early Oligocene–age Florissant shales in Colorado and discovered that the species described by Wickham as *Donacia primaeva* is indistinguishable from the modern leaf beetle species *Plateumaris nitida* (Fig. 4.3). This suggests that *P. nitida* has persisted more than 30 million years.

These and numerous other studies provide strong evidence for the constancy of exoskeletal characters in beetles over great lengths of time. Nevertheless skeptics will ask, "Even if their exoskeletal features show extreme conser-

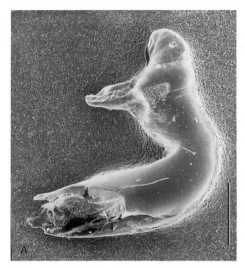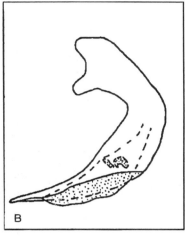

Figure 4.4. Fossil (A) and modern (B) aedeagi of the ground beetle, *Pterostichus brevicornis*. Scale bar in (A) equals 0.4 mm. (Photograph by the author, modern drawing after Lindroth, 1966.)

vatism, how can you be sure that beetle populations have not become genetically differentiated during the Quaternary?" This is a valid question, because paleoentomologists deal only with phenotypic evidence in fossils, not with genotypes.

Three lines of evidence help to resolve this question. First is the physical evidence offered by fossil genitalia. The male aedeagus of many beetles, although an internal organ, is heavily sclerotized (Fig. 4.4). It is therefore resistant to decomposition and is preserved in some fossil assemblages. If a beetle's abdomen is preserved intact, the genitalia are surrounded and protected from physical abrasion in sediments. The study of fossil genitalia has yielded substantive evidence for the constancy of many species through most of the Quaternary and beyond (Coope, 1970). Genitalia are considered the most reliable diagnostic feature in the identification of many beetle species. In fact, there are many beetle species that are so similar externally that their genitalia are the only reliable means of distinguishing between them—they were sibling species then and they are sibling species today.

A second line of evidence for species constancy concerns the stability of the ecological requirements of species. Again, the skeptic may well ask, "How can you be sure that physiological changes have not occurred, causing past populations to have different ecological requirements?" This is another important question, because, if the ecological requirements of a species change through time, it becomes unreliable as proxy data for reconstructing such past conditions as climatic regime, substrates, and host plants. Paleoentomologists are reasonably confident about the constancy of ecological requirements in Quaternary beetles because of the consistency of associations of insect species through time: species kept the same company in the past as they do today.

One of the luxuries afforded the student of abundant, diverse organisms in a fossil record (especially taxa that may be readily identified to the species level) is that many assemblages comprise dozens or even hundreds of species. When the modern habitats and distributions of the species in these assemblages are compiled, a detailed, precise reconstruction of the physical environment and biological community comes into focus. In particular, the overlap in modern distributions of the species in nearly all fossil assemblages corresponds to a fairly narrow climatic "envelope" in which all the species are found to be living (see page 74). If hidden physiological evolution had been taking place in these animals through the Quaternary, there would be species in a fossil assemblage that would be ecologically or climatologically incompatible with the fauna as a whole. This type of discrepancy has rarely been found, though numerous Quaternary insect assemblages have been examined (Coope, 1978, 1979).

In addition, certain well-studied regions, such as the British Isles, have yielded fossil insect assemblages of nearly identical composition during the different glacial, interglacial, and interstadial climatic episodes. In other words, the warm-adapted faunas of one interglacial are extremely similar to the warm-adapted faunas of other interglacials, even though these climatic episodes may have occurred hundreds of thousands of years apart. Likewise, the cold-adapted faunas found during one glacial stage have a great number of species in common with faunas found in previous and subsequent glaciations. If the physiological properties of the species in these faunas were changing through time, then those changes would have to be unidirectional and of constant rate, in order for insect species to assemble in these same communities time after time. The odds of an entire suite of species evolving new physiological requirements in this uniform manner are too small to merit serious consideration.

A third line of evidence for beetle species constancy has been pioneered by Robert Angus. The aquatic scavenger beetle, *Helophorus lapponicus,* was widespread in western Europe during the cold phases of the last glaciation, but its modern distribution reflects the retreat of the species to the remaining cold regions of Europe. These include the northern regions of Scandinavia and mountainous regions as far south as Spain. Angus (1983) captured specimens of *H. lapponicus* from Sweden and Spain; the two populations successfully interbred, despite their genetic isolation during the last 10,000 years.

SPECIES CONSTANCY THEORY

According to Mayr (1970), speciation rate depends on three sets of factors: (1) the frequency of barriers, or factors producing geographic isolation; (2) the rates at which geographical isolates become genetically transformed (how quickly they acquire isolating mechanisms); and (3) the degree of ecological diversity offering vacant ecological niches to newly arising species. When all three conditions mitigate against rapid speciation, an isolate may change hardly at all over millions of years. Angus's data suggest that achievement of genetic isolation requires more than 10,000 years in at least some insects, but Mayr noted that speciation rates vary in different groups of organisms. Cameron (1958) pointed out that the island of Newfoundland, Canada, reinvaded by mammals 12,000 years ago following deglaciation, has already seen 10 of 14 mammal species evolve well-defined subspecies.

The question of number of generations per unit of time appears to have had little to do with species longevity. Mayr (1970) noted that many insects, with one or more generation per year, have not changed in appearance since the Oligocene, whereas slowly maturing mammals, such as proboscideans, evolve very rapidly.

Concerning the longevity of geographic isolation necessary for the development of genetically distinct geographic isolates, Coope (1978) has suggested that insect populations responded to large-scale climatic oscillations in the Quaternary by undergoing shifts in distribution. These shifts were sufficiently frequent to prevent the genetic isolation of populations that would lead to speciation. Many insects are indeed very mobile, a fact that has been inferred from the fossil record and demonstrated in historic times.

Studies of beetle invasion of newly created polders in the Netherlands (Haeck, 1971) and newly created land on the Icelandic island of Surtsey (Lindroth, 1971) demonstrate the rapidity of colonization. For instance, carabid beetles appear to be exceptionally well equipped for rapid migration and establishment in new regions. Many species of carabids contain populations with a mix of fully winged (macropterous) individuals and individuals with reduced flight wings (brachypterous). Studies of ground beetle invasions have shown that flying carabids make up the vanguard of the invaders. Lindroth (1949) described this group of beetles as a "parachute force," capable of spreading out more rapidly at the edges of the range of the species than the flightless "pedestrians." I will discuss this topic more fully in connection with postglacial invasions of deglaciated landscapes in Europe and North America later in this chapter.

Insects invest tremendous amounts of energy and resources in dispersal. Heydemann (1967) estimated that on the North Sea coast of Germany roughly

4.5 billion insects per day are lost by aerial drifting during summer. This is equivalent to 270,000 kg of insect biomass lost to regional insect communities, a high price to pay for dispersal activity.

Present-day ecological studies have shown how existing beetle populations respond to even minor environmental changes. Thus Howden and Scholtz (1986) studied the dung beetle fauna of a wildlife refuge in Texas and found a nearly complete turnover of species during a ten-year interval in which precipitation increased roughly 20%. This suggests that niches do not stay vacant for significant lengths of time, because insect mobility is sufficiently great to allow existing species to occupy new habitats as they become available. An obvious exception to this generalization might obtain on oceanic islands, where invasion of existing species is extremely limited.

It is important to keep in mind that insect habitats are very small and that they are subject to microclimates (but see page 74). Part of the secret of insects' success has been that they have been able to secure suitable habitats for themselves in the face of changing environments by readily dispersing across landscapes. Danks (1979) has summarized many of the factors affecting insect distribution and dispersal. His paper examined the Canadian fauna, but many of the same principles apply to other north-temperate and arctic regions. Disturbances in more or less stable biological communities lead to the development of seral stages of secondary succession. Along with "weedy" plant species, newly disturbed ground is often rapidly colonized by species of beetles and other insects that are adapted for rapid dispersal. Such species are among the more important pests of agricultural crops. These species are opportunistic, with rapid reproduction and great mobility. Their habit of feeding on the flora of disturbed ground has allowed them to take advantage of agricultural habitats created within the last few hundred or thousand years. In the Pleistocene, glacial advances and retreats created newly exposed open ground habitats in many regions. The adults of many of these insect species go through a pre-reproductive dispersal phase. Though many of the dispersing individuals perish, the colonization of newly disturbed open ground habitats is assured for the species as a whole. Vagility, or an intrinsic tendency to disperse, is a reflection of the inherent instability of landscapes.

Although there is good evidence for long-distance dispersal in some insects, some groups seem to disperse very little. Lindroth (1957) showed that many of the beetle species accidentally imported into the east coast of North America in ship ballast have remained in the immediate vicinity of their ports of arrival. This restricted dispersal may be due to the species' lack of ability to compete with the indigenous fauna. Competition between species is an important factor in limiting distributions, but for most groups of organisms it has thus far been little studied.

Insects exhibit many types of distribution patterns. Some species, though widely distributed, occur only in narrowly defined habitats within their range (stenotopic species). For instance, there are several species of beetles that occur today across all of northern North America, but live only in bogs, such as the rove beetle, *Gymnusa atra*. Other species are found in a wide range of habitats (eurytopic species), but are confined to a narrow geographic range. Still other species are cosmopolitan in both their distribution and habitat requirements.

In the fossil record, we see dramatic evidence of shifts in insect ranges through Quaternary time. The story of the Tibetan dung beetle *Aphodius holdereri* in British Pleistocene assemblages reveals the magnitude of possible distributional change in Eurasia (Coope, 1973). This type of change has also been shown in the water scavenger beetle, *Helophorus mongoliensis*, a modern inhabitant of Asiatic mountains that was found in 40,000-year-old deposits in Britain (Angus, 1973), and in the rove beetle *Tachinus caelatus*, known today from Mongolia but found in deposits from the last glaciation in Britain (Taylor and Coope, 1985) and Switzerland (Coope and Elias, unpublished data).

Micropeplus hoogendorni was originally found in Alaska in Pliocene deposits, but its discovery in mid-Pleistocene deposits from Britain and in the modern fauna of Siberia shows that interhemispheric distributional shifts have also occurred. In addition, the rove beetle *Anotylus gibbulus* has been found in Sangamon or early Wisconsin–age assemblages from the Scarborough Bluffs in Toronto (Hammond et al., 1979), and rarely from interstadial deposits in the Devensian (last) glaciation in Britain. *A. gibbulus* first occurs in interglacial deposits in Britain of 200,000 years ago, when it was the most abundant rove beetle. This beetle has a most peculiar distribution today; it is known only from the Caucasian Mountains and possibly from eastern Siberia north of Vladivostok. These species clearly have been world travelers during the Quaternary. The biogeographic implications of the fossil record are discussed in Chapter 6, but for now it is sufficient to note the great mobility of insects in response to environmental change.

Kavanaugh (1979) has suggested that the beetle species' longevity as evidenced in the Quaternary fossil record applies only to lowland taxa, since nearly all fossil studies have focused on lowland regions. In his analysis of speciation in the ground beetle genus *Nebria*, Kavanaugh proposed the differentiation of 26 subspecies pairs and three species pairs in post-Wisconsin time (i.e., within the last 10,000 years) and the differentiation of another 10 *Nebria* species pairs earlier in the Quaternary period. In particular, he discussed the *trifaria* species group as showing very recent speciation, arguing that isolation in montane habitats brings about rapid speciation. However, the fossil record suggests that some beetle species associated with mountaintop habitats today exhibited considerable vagility in the past. Late Wisconsin–age insect fossil assemblages from Lamb Spring, in the foothills east of the Rocky Mountains of Colorado, contained several species in this distributional category that apparently shifted downslope to the plains during the last glaciation (Elias, 1986; Elias and Nelson, 1989; Elias and Toolin, 1989). One example of this finding is provided by the water scavenger beetle, *Helophorus splendidus,* which is found today in the arctic and on the alpine tundra in Colorado (Smetana, 1985). It occurs in a 14,500-yr B.P. assemblage from Lamb Spring. It was very common in England prior to 13,000-yr B.P.

Coope (1979) argued that rapid speciation requires environmental stability in a constant geographic location. It is tempting to speculate that the tropics might offer such stability, but even tropical regions seem to exhibit significant environmental fluxes (Colinvaux, 1987; Markgraf and Bradbury, 1982). Erwin (1979) hypothesized that changes in Pleistocene precipitation patterns in tropical regions caused the last major impulse of ground beetle evolution. On the other hand, the remarkable diversity of insect life in the tropics may be due at least in part to a lack of intensity of environmental change over long periods.

Coope (1979) also suggested that isolated geographic situations (i.e., habitat islands) such as equatorial mountaintops, oceanic islands, and caves may represent evolutionary traps from which emigration is denied. Under such conditions, rapid evolution would be the only viable response of a species to environmental change. However, some beetle species appear to be relatively conservative, even on oceanic islands. On Aldabra, a recent atoll in the Indian Ocean, Basilewsky (1970) found 18 species of ground beetles, none of them endemic. All of these species have been collected on Madagascar, and many are also known from the continent of Africa (Thiele, 1977). On the Galapagos Islands, carabid beetles have had roughly 10 million years to evolve, although surveys of carabids on the islands have found surprisingly few species, and Thiele (1977, p. 306) summarizes carabid speciation on the Galapagos Islands as "not very impressive" in comparison with that of other groups of animals.

Cave faunas also probably are not as genetically isolated as might be expected. My recent studies of late Quaternary insect fossils from packrat middens in the Chihuahuan Desert (summarized in Elias, 1992b) included the discovery of several species of cave beetles in the ground beetle genus *Rhadine* that have shifted their distributions in response to environmental changes in the last 20,000 years. These beetles are flightless and depigmented; most have reduced eyes. Their fossil record indicates that they have somehow changed their ranges from caves in the Chihuahuan Desert to caves in central Texas and Oklahoma, even though no subterranean connections are known between these far-distant cave systems. Barr (1960) discusses the possibility of geographic isolation of ancestral stocks of these cave beetles in west Texas during the late

Pliocene or early Pleistocene, but the fossil evidence indicates that these species are able to shift substantial distances within at most a few thousand years. In light of these findings, fossil insect studies should also be carried out on oceanic islands and equatorial mountaintops in order to test their viability as evolutionary traps.

Bennett (1990) discussed the effects of Milankovitch cycles on Quaternary biota. Variations in the earth's orbit bring about climatic changes on frequencies varying from 20,000 to 100,000 years. Bennett argues that recent evolutionary theories, such as the punctuated equilibrium theory (Gould and Eldredge, 1977), overlook the significance of climatic changes on the 20,000- to 100,000-year scale. Several time scales are at work in the evolutionary history of a species. One of these is the generation time, which for most insects is one year or less. A second time scale is the duration of species, which the fossil record now suggests may be up to several million years, at least in Coleoptera. Stanley (1985) surveyed species duration in a wide variety of organisms and concluded that species endure for periods of from 1 to 30 million years, depending on taxonomic group. Species therefore persist much longer than Milankovitch cycles, even though these cycles are thought to have brought massive climatic changes in the Quaternary. These environmental changes have caused the essentially constant disruption of biotic communities for more than two million years. Each species has reacted individually to these changes. Again, the fossil record suggests that insects, as well as many other organisms, have shifted their distributions on regional, continental, and even intercontinental scales in the Quaternary. Bennett argued that macroevolution could scarcely proceed by phyletic gradualism in light of this constant shifting and disruption of populations. Coope (1970) anticipated the punctuated equilibrium theory in his discussion of speciation of beetles in the Quaternary.

Bennett also noted that most paleontological research deals with time scales of millions or tens of millions of years, too coarse to resolve events at finer intervals. At the other extreme, modern ecological research deals with events occurring over only a few years at most. But the really important evolutionary events fall in between these two scales. Quaternary paleontology fills this gap precisely. More detailed study of Quaternary fossil records will illuminate the history of life on time scales between the paleontological and ecological extremes. The type of change that has occurred in the Quaternary may well have been operating throughout the history of life on earth.

Study of the Quaternary fossil record first began to reveal startling data on the longevity of beetle species about twenty years ago. Unfortunately, these studies have largely gone unnoticed by some beetle systematists and evolutionary biologists. Futuyma (1979) cited a study claiming that poorly flying bog beetles have failed to migrate north of the Wisconsin ice limits in North

America since the end of the last glaciation. Yet the fossil record indicates that even flightless species have migrated across major regions of North America and that these distribution shifts have taken only a few centuries at most (Elias, 1991). This is the type of information that needs to be disseminated outside of the field of paleontology, so that its scientific connections with evolutionary biologists may be strengthened. Interactions between the two fields will be mutually beneficial.

In summary, the fossil record provides evidence for both the longevity of species and the almost constant perturbation of biotic communities and shifting of species' distributions in response to environmental change during the Quaternary. These features of life on a 20,000- to 100,000-year time scale may lend support to the punctuated equilibrium model of evolution, but we are nevertheless left with some major questions. As Ashworth and Hoganson (1987) pointed out, the problem of trying to assess the precise antiquity of species is that the evidence is inadequate to the task. The fossil data (at least those for the late Quaternary assemblages) are adequately dated but lack the precision needed to define close evolutionary relationships. On the other hand, cladistic analysis may succeed in establishing evolutionary relationships but lacks real temporal control. Only recent fossils can provide such a temporal framework.

SENSITIVITY OF INSECTS TO ENVIRONMENTAL CHANGE

One of the most important characteristics of insects as paleoenvironmental indicators is their sensitivity to environmental change. Species in many families of beetles have demonstrated such sensitivity, and this is also true for caddisflies, midges, and other insect groups. In general, predators and scavengers receive the most attention in paleoenvironmental reconstructions because they are able to respond more rapidly to climate change, since they are not tied to specific types of vegetation. Whereas some predators and scavengers are eurythermic (adapted to a broad range of thermal conditions), many are stenotherms that are adapted to only a narrow thermal environment. Stenotherms may rapidly colonize a region as long as the climatic conditions are suitable. When climatic conditions change stenotherms disappear with equal rapidity.

Insect ecologists support the idea that insect abundance and diversity may be controlled by biotic factors (e.g., predation, competition, or parasitism) in the center of the range of a species, but that abiotic factors such as climate probably limit populations toward the edges of the range (Price, 1984). It is best, therefore, to study fossil assemblages from ecotones (the edges of ecosystems) rather than assemblages from the centers of past ecosystems. Another advantage to studying the ancient biota of ecotones is that these communities most readily document regional environmental changes. Fossil faunas and floras change as ecotonal boundaries shift across a landscape. These biota are quite literally on the leading edge of change. On the other hand, communities from the center of ecosystems tend to be complacent, because ecotonal boundaries seldom move through these regions.

A great deal of experimental work on the habitat preferences and physiology of beetles has focused on the well-studied family Carabidae. This is due in no little part to the pioneering work of Carl Lindroth (1949–1987) on the ground beetles of Scandinavia.

Thiele (1977) determined the thermal preferences of ground beetles, using controlled experimental environments that provided a gradient of temperatures to the specimens. These tests established a preferred temperature (PT) for a large number of European species. Thiele found no evidence for "temperature races" in carabids (i.e., populations within a species that have differing PTs). In addition, his experiments showed that stenothermic species associated with arctic and alpine environments, such as *Nebria nivalis* and *N. gyllenhalli,* exhibited PTs of 5° and 8°C, respectively. At the other end of the thermal spectrum, only one species of ground beetle from western Europe, *Callistus lunatus,* was found to have a PT above 40°C. These results offer substantial evidence linking the geographic distribution of ground beetles to their thermal requirements.

Additional experiments on the metabolic rates of carabids (Thiele, 1977) showed that beetles' optimal metabolisms correspond with their PTs. This phenomenon intensifies during the process of cold-hardening, which takes place in late summer or fall. In preparation for the onset of winter cold, ground beetles in northern regions become progressively cold-adapted (able to function at cold temperatures). Under these conditions, the beetles' metabolic rates peak at lower temperatures, and they may become paralyzed or die if exposed to the temperatures they experience in midsummer. Circadian rhythm in arctic species is a residuum from their time (normal) spent further south.

Surprisingly, terrestrial species are often the most abundant taxa in water-lain sediments. On the whole, they probably reflect regional macroclimates better than aquatic species. However, many aquatic insects (including aquatic beetles, caddisfly larvae, and chironomid larvae) are adapted to waters in a narrow range of temperatures. These water bodies are developed and maintained only within certain climates, so they indirectly reflect macroclimate. Williams et al. (1993) compared fossil caddisfly and beetle paleoclimate reconstructions for sites in the Great Lakes region and concluded that the two groups respond nearly identically to climate change.

Aquatic insects also provide a basis for inferring detailed limnological data, including types of substrates, water pH, water clarity versus turbidity, and trophic status. Riffle beetles (Elmidae) provide evidence of current speed in streams.

Phytophagous beetles may be less useful in paleoclimate reconstructions because they may merely reflect host plant distribution, but they still provide data valuable in paleoecological studies. They are reliable indicators of past vegetation, and host plant–specific beetles occur in many assemblages (although bark beetles are found only in forest assemblages). Bark beetles and other tree-associated groups (e.g., some weevils and twig borers) have been used to document ecological changes in forests and to estimate the position of both latitudinal and altitudinal treeline. Fossil scolytids from Holocene peats at Ennadai Lake, Keewatin, Canada, indicated the continued presence of spruces at the site, even though the mid-Holocene pollen record was lacking in spruce pollen (Elias, 1982a). I have also used changes in the ratio of bark beetles to alpine tundra beetles to infer the history of altitudinal treeline in the Rocky Mountains (Elias, 1983, 1985, 1988b; Elias et al., 1991).

In Britain, the presence of fossils of the elm bark beetle, *Scolytus scolytus,* occurred at the same time as the mid-Holocene elm decline. The cause of the elm decline is difficult to unravel because it may have been completely natural, or it may have been compounded by the activities of early Neolithic people. Whatever the cause, the decline may have been exacerbated by the spread of Dutch elm disease, which is transmitted by the beetle (Girling, 1988).

Besides predators and phytophages, other groups of beetles and other insects can play an important role in reconstructing past environments. Fungus beetles provide part of the ecological story of past communities. Many species are associated only with particular types of fungi, such as bracket fungi or mushrooms that grow only in old, well-established stands of trees.

In archaeological studies, the presence of dermestid beetles (Dermestidae) and other stored product pests is used to infer the nature and patterns of human food use and sanitation (or the lack of it) (Osborne, 1977). The beetles that have become successful as stored product pests obviously did not start out in this niche in the late Tertiary or early Quaternary; however, Buckland (1981a) noted that human populations have played a major role in their recent dispersal. There is some speculation that these species previously fed on caches of food stored by other animals, such as squirrels and other rodents (Crowson, 1981).

Fossils of parasitic insects are used to infer the presence of domesticated animals (livestock) in archaeological sites. Even if no cow or sheep remains are preserved in an ancient dwelling, the exoskeletons of the animals' parasites may provide the data needed to infer the practice of animal husbandry (Buck-

land, 1976a). Fossil remains of human parasites (fleas, lice, and other organisms) document the history of human life-styles and associated health
problems, because insects are associated with human beings from cradle to
grave. Various insect genera document the forensic pathology of corpses, from
desiccated Greenland mummies (Bresciani et al., 1983, 1989) to the victims of
ritual murders in Iron Age Britain (Girling, 1986).

As in other macrofossil studies, the "story" obtained from a fossil insect
assemblage has particularly local application. A single local reconstruction
may not provide a reliable estimate of regional environments. For instance,
Ashworth (1977) described a late Wisconsin beetle fauna from southern Ontario that was indicative of a cold microclimate in proximity to stagnant ice, but
additional regional studies from this period clearly show regional climatic
amelioration (Morgan et al., 1984a). Therefore, to develop an understanding of
a region, numerous replicates of local studies are required.

Insects are useful to paleoecological reconstructions because they are such
important elements in terrestrial ecosystems. They are the most abundant and
diverse group of animals, as they have been since before the age of dinosaurs.
Insect species constancy throughout much of (if not more than) the last million
years allows us to make use of ecological and distributional data drawn from
modern populations. As such, their fossil record provides very detailed, precise
information on such factors as vegetation, soils, water quality, forest composition and health, and even vertebrate species composition (from dung beetles
and a variety of host-specific arthropod parasites). Predators and scavengers
supply reliable information on past climates. When taken together, the biotic
and abiotic elements reconstructed from fossil insect assemblages allow the
investigator to bring an ancient community into sharp focus.

5`

PALEOCLIMATIC STUDIES
USING INSECTS

The transitions from one climatic mode to the other were so sudden at times
that the whole terrestrial biota was disrupted.
—Russell Coope (1987a)

Given that the species found in Quaternary fossil assemblages had for the most
part the same environmental requirements as their modern counterparts, it is
possible to construct a mosaic of environmental conditions for given study sites
and time intervals, and then to weave these environmental reconstructions into
a regional synthesis depicting macroclimate. The presence of suitable climatic
conditions is one of the most important factors determining the geographic
range of insects (Coope, 1986).

Starting with Coope's (1959) interpretation of late Quaternary environments at the Chelford site, most climatic interpretations of fossil beetle
assemblages have been made on the basis of information gleaned from
average temperature and precipitation data from meteorological stations
that lie in the modern ranges of the species in a given fossil assemblage. The
approach taken in most studies has been to plot the region in which the
modern distributions of the species in an assemblage overlap, and then to
derive a paleoclimate estimate based on the modern climatic parameters
within that zone of overlap.

Although this technique, known as the range overlap method, has generally
yielded good results, it is not always viable. For instance, during intervals of
rapid climate change, insect faunas (as well as the rest of the regional biota) are
in such a state of flux that their presence in any one region may be very
ephemeral. Fossil assemblages representing such faunas often contain mixtures
of species for which there is no modern analogue. By that I mean that there is
no one region in existence today where all the species represented in the fossil
assemblage can be found living.

The range overlap method is, to a certain extent, subjective. Investigators are
called upon to exercise their best judgment regarding distributional or climatic
data that appear out of place relative to the majority of the evidence. Species

deemed to be especially climatically sensitive (stenothermic) become "indicator species" in an assemblage, and hence receive more weight in paleoclimatic reconstructions than their more cosmopolitan (eurythermic) counterparts. In the 1980s, the mutual climatic range method was developed to standardize paleoclimatic interpretations of fossil beetle assemblages and to tease more climatic information out of the data.

THE MUTUAL CLIMATIC RANGE METHOD

In 1982, Russell Coope sought the help of paleoclimate modelers Timothy Atkinson and Keith Briffa in the Climatic Research Unit at the University of East Anglia in developing a quantitative method for analyzing the paleoclimatic interpretations of Quaternary insect fossil data. The method they devised, termed the mutual climatic range method (MCR), avoids the use of indicator species, concentrating instead on the analysis of entire assemblages, using the presence or absence of species rather than their relative abundance (which may vary considerably, depending on depositional environments). The basic principle of the method lies in establishing the range of climates presently occupied by each beetle species found in a fossil assemblage. A species need not necessarily occupy the whole of its potential range, nor need we have a complete picture of its geographic distribution. In order for an adequate climatic envelope to be constructed, all that is needed is a determination of the climatic tolerances of a species, based on the climatic parameters within its known range. This technique thus presents a great advantage over the geographical overlay method in that it recognizes that species may live in identical physical environments but in different places.

Once the climatic envelope for each species has been established, the climate indicated by the whole assemblage may be assumed to fall within the area of overlap of the climatic ranges of all the species in the assemblage (Atkinson et al., 1986). Beetles are especially well suited to this technique because they are a varied group in which many species show clearly defined thermal tolerances (Atkinson et al., 1987). On a hemispheric scale, the distribution of individual species reflects temperature regime, especially summer warmth and degree of seasonality (i.e., annual temperature range).

Atkinson et al. (1986) have tested the MCR method on modern beetle assemblages from Europe and Asia, and the results of the climate reconstructions match the modern climates associated with the modern assemblages studied. The following description of the method is based on their account.

The first step is the compilation of modern distribution maps for each species. For many species, adequate modern distribution maps have been published. For others, however, additional literature searches and transcribing of locality labels from museum specimens will have to be performed. Only predators and scavengers are used. Focusing on these groups avoids the problem of dealing with phytophagous species, the distributions of which may reflect their host plants' ranges more than climatic parameters. In contrast, predators and scavengers are more or less free to become established in any region of suitable climate, even as part of pioneering communities establishing the first "foothold" on bare, mineral soil in recently deglaciated landscapes.

The second step is the determination of climatic range for each species. The geographic range of each species is converted into a climatic range with the help of a base map showing locations of meteorological stations. A comparison of the two maps reveals whether given meteorological stations are within a species' geographic range or outside of it. Then the climate data from the two groups of meteorological stations are processed by computer, producing a plot of the two groups of stations on a graph. Principal components analysis of the mean monthly temperature from 495 meteorological stations in the Palearctic region shows that over 96% of the variance in temperature regime for the Palearctic is described by two groups of variables, which can be interpreted as (1) summer warmth (TMAX) and (2) temperature range between the warmest and coldest months (TRANGE) (Atkinson et al., 1987). The latter serves as an index of degree of seasonality. The range of climates in which the various species could coexist is reflected by the overlap between the climatic envelopes, as shown for two hypothetical species (Fig. 5.1). It is also possible to estimate the mean temperature of the coldest month (TMIN) by constructing diagonal isolines and again taking the median value and extremes covered by the mutual overlap.

For species with modern distributions that include montane sites, care must be taken to match the altitudinal distribution of the species with that of the meteorological stations. The montane species problem illustrates why it is important to find climate data as far as is possible from the nearest meteorological station, and not derive modern climate parameters on the basis of the position of modern beetle collecting localities relative to broad isotherm lines, such as are found in regional climatic atlases. Regional climatic summaries expressed in isotherms may fail to show altitudinal gradients in temperature values, simply because the data are too complex to represent graphically on colored isotherm maps of large regions.

The plotting of species climatic envelopes groups together widely separated geographic locations where a beetle species occurs under similar

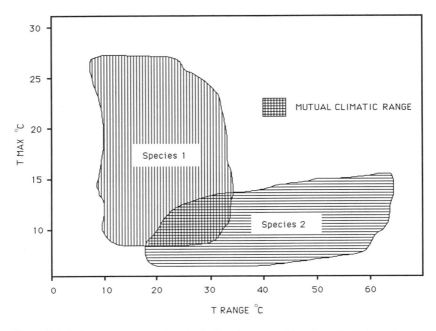

Figure 5.1. Tolerance ranges and mutual climatic range of two hypothetical beetle species. (After Atkinson et al., 1986.)

climates and reveals the climatic homogeneity underlying many species' distributions. A widely scattered data set condenses into a compact and more easily handled unit.

The third step in the MCR process is computer storage and retrieval of species climatic envelopes. In order to facilitate rapid calculation of paleo-temperatures by computer, the climatic range envelope for each species is coded and stored in numeric form. This is achieved by superimposing a grid (36 × 60 in 1°C units) on the TMAX/TRANGE plot and designating each element by a 1 or a 0, according to whether it is within or outside the envelope.

The fourth step is climatic reconstruction from an assemblage of named species. Given a list of species in an assemblage, the computer retrieves and superimposes their numeric envelopes to produce a TMAX/TRANGE graph of percentage overlap. Visual inspection of this plot allows the area of maximum overlap to be determined. Then the values of TMAX, TRANGE, and TMIN are taken from that area.

The MCR method lends itself to rigorous checks by reconstructing modern climates from modern beetle faunas collected within a restricted area and

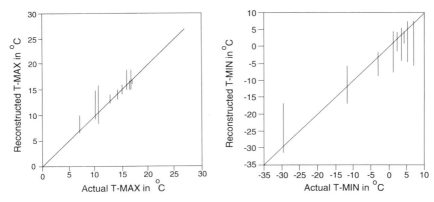

Figure 5.2. Test of the MCR method on assemblages from Europe, Iceland, and Siberia. Reconstructed temperatures of the warmest (TMAX) and coldest (TMIN) months are shown on the vertical axes; actual TMAX and TMIN values are shown on the horizontal axes. (Data from Atkinson et al., 1987.)

comparing the results with the mean temperature records from nearby meteorological stations. The results of a test based on species living today at 15 localities in Europe, Iceland, and Siberia are shown in Fig. 5.2. For the coldest locality studied (Chaun Bay, Siberia), the values of TMAX and TMIN had to be estimated from a climate atlas, in the absence of any nearby meteorological data.

This test and other tests suggest that the accuracy of the method is acceptable, using current distributional and meteorological data bases. Atkinson et al. (1987) have also published a more refined calibration method, including a correction for overestimation of TMIN for colder climates and its underestimation in climates with mild winters. These systematic errors are corrected using regression equations of modern temperatures at various localities against the median reconstructed temperatures based on the beetle fauna from the same localities. The regression equations are as follows:

$$\text{TMAX(corrected)} = [1.006\text{TMAX(median)}] + [0.0142 \text{ NSPEC}] - 2.96$$
$$(r = 0.94; s = 0.83°C)$$
$$\text{TMIN(corrected)} = [1.416\text{TMIN(median)}] + 1.904$$
$$(r = 0.94; s = 2.42°C)$$

where temperatures are in degrees Celsius and NSPEC is the number of species used in the reconstruction. Corrected values obtained using these equations provide unbiased estimates of the most probable paleoclimate within the MCR, with a precision of about ±2°C for TMAXD and about ±5°C

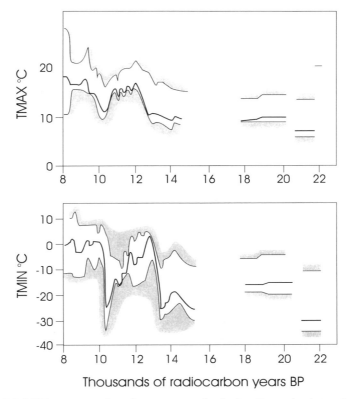

Figure 5.3. MCR reconstruction of temperatures for the late Devensian interval in the British Isles. Shaded area defines the limits of the MCR; bold line shows the most probable value of paleotemperatures. (After Atkinson et al., 1987.)

for TMIN. With further additions and refinements the accuracy of the method will be improved.

An actual example of the application of the MCR method is shown in Fig. 5.3. The values of TMAX and TMIN are based on time-averaged data. The shaded upper and lower boundaries define the limits of the MCR (paleotemperature estimates lie within these limits). However, the bold lines show the most probable values of paleotemperatures as determined by the correction equations given earlier. Gaps represent periods for which data were not available. The MCR reconstructions for Britain are remarkable in that they show the rapidity and strength of climatic change at the end of the last glaciation. Extremely rapid warming took place at about 13,000 yr B.P., and again at 10,000 yr B.P. In between, the marked cooling associated with the Younger Dryas is also demonstrated.

In summary, the MCR method devised by Atkinson et al. (1986, 1987) has been shown to be an extremely useful tool for paleoclimate reconstructions. Fossil beetle proxy data were already providing some of the best paleotemperature data, because beetles are sensitive, reliable indicators of thermal regime. MCR analyses of faunas previously interpreted using the range overlap method have seldom altered those interpretations in any significant way. The paleoclimate curve developed by Atkinson et al. (1987) differs little from an earlier curve drawn by Coope (1977) without the benefit of MCR analysis. However, the addition of the MCR method allows the proxy data from beetle fossils to be quantified, calibrated, and tested against those for modern communities. MCR also permits estimates of winter temperatures and degree of seasonality.

I have recently received a grant to develop MCR analyses of late glacial insect fossil data from a transect of sites in North America. This will be the first time that the MCR method has been employed outside Europe.

INSECT EVIDENCE FOR RAPID CLIMATE CHANGE

The previous chapters have shown that insects are in many ways uniquely suited as sources of proxy data for the reconstruction of past terrestrial environments, both on land and in fresh water. They are sensitive to environmental change, and they respond to such changes in ways that can be indicated through fossil assemblages. These phenomena are amply demonstrated in Quaternary insect records.

One of the most important contributions that insect fossil data have made to our understanding of Quaternary environments is the evidence they provide for rapid, intense climate changes. Prior to Coope's work in Britain, nearly all terrestrial paleoclimatic reconstructions were based on studies of past vegetation, interpreted from pollen. Many palynologists assumed that plant communities were reliable indicators of climate change and that shifts in vegetation patterns were synchronous with climatic fluctuations. Most changes registered in pollen spectra through the Quaternary have been gradual, with transitions between glacial and interglacial episodes lasting hundreds if not thousands of years. However, Coope's work began to cast doubt on paleobotanical reconstructions of climate during intervals of rapid change, especially during the late glacial interval. In fossil insect records, changes between faunas suggestive of major climatic episodes may occur in as little as a few decades, and are often so rapid as to appear instantaneous in the fossil record. These data suggest that climatic changes would be graphed as an almost square wave pattern, as opposed to the gentle, sinusoidal curve of climatic change as interpreted from pollen data.

Evidence from British Studies

Let us now examine two episodes of rapid, intense climate change first inferred through analysis of fossil insect assemblages in Britain. One is a rapid warming (the Upton Warren Interstadial Complex) in the middle of the last glaciation. The other is the series of events at the end of the last glaciation.

The beginning of the Upton Warren Interstadial Complex is marked by an almost instantaneous replacement of an arctic-subarctic beetle fauna with a temperate fauna (Coope et al., 1961; Coope, 1977). The timing and exact rapidity of this climatic change remain unknown, because it would appear to have taken place about 43,000 years ago, which places it at the limit of reliable radiocarbon dating. The age of the arctic fauna (44,300 ± 1600 yr B.P.) overlaps the age of the temperate fauna (43,000 ± 1200 yr B.P.). Stratigraphically, however, the arctic and temperate faunas are separated by only a few centimeters of sediment (Coope, 1977). The British pollen records from this interval continued to indicate tundra conditions associated with scant, herbaceous vegetation. Even though the insect evidence from this warm episode indicates conditions sufficiently warm to allow the establishment of trees in Britain, evidence is lacking the the trees arrived before the subsequent climatic deterioration leading to the last glaciation.

The late glacial sequence from Britain (Fig. 5.3) is better understood than the Upton Warren Interstadial Complex, mainly because late glacial events (14,000–10,000 yr B.P.) have been reliably dated by radiocarbon assay. Insect data indicate that the British late glacial warming was sudden and intense, beginning 13,000 years ago. It is signaled by a replacement of arctic and subarctic species by temperate species. Beetle evidence points to a rise in summer temperatures of 7°C and a rise of winter temperatures of perhaps as much as 20°C. These changes constitute a replacement of cold, continental climate by mild, oceanic climate within the space of a few decades (Coope and Brophy, 1972).

Following the Younger Dryas oscillation, warming of similar magnitude and rapidity took place (Osborne, 1974; Bishop and Coope, 1977). This reconstruction of climatic changes is considerably different from the traditional view (based on palynological evidence) of European climatic changes at the close of the last glaciation (Coope, 1987a). The pollen record from the early phase of the late glacial in northwestern Europe shows a continuity with vegetation from the previous glacial interval. As Coope and Brophy (1972) stated, the palynological interpretation of tundra conditions during the early phase of the late glacial is based largely on negative evidence, that is, on an impoverished flora and a lack of trees. But the vegetation from this time is composed largely of pioneering plant species. Their presence should be viewed as evidence for a lack of competition in an impoverished flora, rather than as an index of arctic conditions.

Independent paleoclimatic reconstructions based on oxygen isotope ratios in Greenland ice cores offer corroboration of the British paleoclimate scenario based on insect data. Dansgaard et al. (1989) inferred climatic warming of 7°C within 50 years after the termination of Younger Dryas cooling at 10,700 yr B.P. This is precisely the same estimate of amelioration that Coope made for terrestrial environments in Britain at this time. In summarizing oxygen isotope records from the Greenland ice sheet, Dansgaard (1987) noted that the isotope fluctuations in the late Quaternary represent a series of abrupt and drastic changes in the North Atlantic environment.

Moore compared British paleoclimatic reconstructions based on mammals with those based on insects and concluded that "beetles are better climatic indicators than bears" (1986, p. 385). Two problems are inherent in the mammalian fossil record. One is the paucity of specimens in comparison with those of invertebrates, pollen, and diatoms. A second problem is that climatic interpretations based on mammalian fossils are complicated by the relatively large range of climatic tolerance in mammals, compared with the narrower climatic tolerance of some poikilothermic invertebrates. In comparing the British fossil record of the last 120,000 years, Moore concluded that "perhaps the beetles will still have the last word" (1986, p. 386).

The British beetle evidence argues strongly that vegetational change lagged behind the rapid changes in climate in the late Pleistocene. This lag is most likely due to the slow migration rates of plants, especially trees—a topic that will be discussed more fully later in this chapter. If this phenomenon were only seen in the British fossil record, it might be dismissed by many as an accident of biological history, due perhaps to the unique geographic position of the British Isles. Indeed, were it not for the warming effects of the Gulf Stream, the northerly position of much of Great Britain would dictate climatic conditions similar to those in central Labrador. The shifting of the Gulf Stream's position in the North Atlantic may have played a key role in the speed and intensity of climatic changes in Britain during the late glacial (Ruddiman and McIntyre, 1981).

Evidence from Continental Europe

In fact results of studies from a wide variety of landscapes confirm the British evidence quite convincingly. Numerous episodes of rapid change

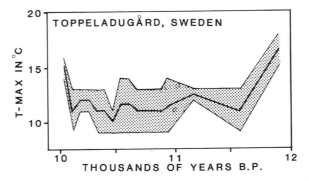

Figure 5.4. MCR reconstruction of summer temperatures for the late glacial period at the Toppeladugård site, Sweden. Shaded area defines the limits of the MCR; bold line shows the midpoint of each MCR estimate. (After Lemdahl, 1991a.)

indicated by the Quaternary insect record could be discussed here, but the most fully documented one is the late glacial to Holocene transition. This interval is the best-dated episode of rapid climate change in the late Quaternary, and organic deposits of this age are found in many regions; thus it is the most thoroughly studied of the major Quaternary climatic transitions.

In Sweden, Lemdahl (1985, 1988c, 1991a) documented late glacial climatic changes from a series of insect fossil assemblages in Scania. These faunas show an initial late glacial amelioration slightly later than that seen in Britain (circa 12,600 yr B.P. versus 13,000 yr B.P.) but of a similar intensity (Fig. 5.4). In contrast, the Swedish vegetation shows a gradual change in the early part of the late glacial, also known as the Bølling pollen zone. Even though the insect fauna is comprised of species from the boreal zone, the regional vegetation was open ground tundra with shrub birch (Iversen, 1954; Berglund et al., 1984). Birch forest arrived in southern Sweden during the Allerød pollen zone, nearly a thousand years after the initial climatic warming.

Following a Younger Dryas cooling, another rapid warming began at about 10,200 yr B.P. The temperate vegetation signature of the Preboreal pollen zone lagged behind this amelioration by about 500 years (Lemdahl, 1985, 1991a).

Lemdahl (1991b) described a late glacial insect faunal sequence from Zabinko in western Poland that indicates the arrival of temperate insect species as much as 500 years earlier than in southern Sweden. He suggested that this differential was due to the impact on Swedish climates of the slowly retreating Scandinavian ice sheet, which departed western Poland much earlier (circa 18,000 yr B.P.). Thus the initial amelioration in Poland was synchronous with

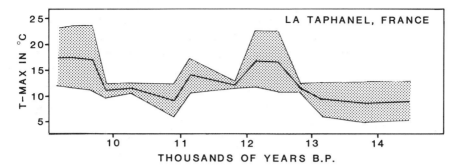

Figure 5.5. MCR reconstruction of summer temperatures for the late glacial period at La Taphanel, France. Shaded area defines the limits of the MCR; bold line shows the midpoint of each MCR estimate. (After Ponel and Coope, 1990.)

the amelioration in Britain. Even though the insect evidence suggests summer temperatures of 14–15°C, subarctic vegetation was associated with the initial warming signaled by the beetle fauna (Tobolski, 1988).

Ponel and Coope (1990) studied a late glacial insect faunal sequence from La Taphanel at an altitude of 975 m in the Massif Central region of southern France (Fig. 5.5). The timing and intensity of climatic changes were very similar to the British reconstruction, with rapid amelioration by 13,000 yr B.P., a marked Younger Dryas cooling just after 11,000 yr B.P., and another rapid warming before 10,000 yr B.P. Interestingly, the reconstruction of climatic events at La Taphanel based on insect data is much more synchronous with the vegetation record than that for northern Europe. Presumably, this is because the temperate plant species that survived the last glaciation in refugia in southern Europe were able to migrate rapidly into the Massif Central region when climate ameliorated. However, a brief climatic cooling at about 12,000 yr B.P. (corresponding to a synchronous cooling in Britain) was inferred from the insect fauna but was lacking from the pollen record (de Beaulieu et al., 1982, 1984). Even when plant migration routes are short, it appears that some insect groups (e.g., ground beetles) that are commonly found in fossil assemblages do a better job of signaling rapid or short-term climate change than most plants represented in fossil pollen spectra.

In Switzerland, late glacial insect assemblages have been described from Lobsigensee on the Swiss Plateau (Elias and Wilkinson, 1983) and from Champreveyres, adjacent to the Jura Mountains at the Lake of Neuchâtel (Coope and Elias, unpublished data). The Lobsigensee record contains an abrupt change from an arctic and alpine fauna to a temperate, boreal fauna at about 13,000 yr B.P. As at La

Taphanel, the vegetation and insect records from Lobsigensee appear to be synchronous at the beginning of the late glacial warming. The shift from arctic and alpine to temperate insect assemblages coincides with the evidence from pollen spectra, including tree birch, willow, and juniper in the Bølling pollen zone (Ammann et al., 1983). However, whereas the insect fauna in the Bølling is indicative of climatic conditions found today in central Europe, available evidence suggests that the modern vegetation of that region was not developed at Lobsigensee until 2500 years later. This pattern is often repeated in late glacial and early Holocene paleoecology: insect faunas suggest an abrupt, early warming to levels very near modern parameters, and vegetation appears to develop slowly from open ground through shrubs to pioneering trees (different taxa in different regions), finally establishing the more or less modern plant communities.

The earliest insect faunal assemblage from the Champreveyres site (Coope and Elias, unpublished data) has been dated at about 12,800 yr B.P. This assemblage reflects temperate conditions, so the timing of the transition from late glacial climates predates it. The presence of thermophilous beetles in early late glacial assemblages suggests that the climate was warm enough to have sustained mixed deciduous forest by at least 12,800 yr B.P., although regional forests of modern composition seem not to have been established until the Holocene, more than 2000 years later. Climatic warming alone is not sufficient to induce forest establishment. All of the proper conditions must first be met, including appropriate soil chemistry and organic content, seasonality, and quality of moisture.

Evidence from North America

Morgan et al. (1984a) summarized some of the work that has been done on late glacial insect fossils in North America. The most intensively studied region includes the central and eastern United States and southeastern Canada. A series of insect faunas indicate changing conditions at the end of the last (Wisconsin) glaciation. The Longswamp site in southeastern Pennsylvania (Morgan et al., 1982) contained a basal fauna indicative of conditions immediately postdating the maximum extent of the Laurentide ice sheet, at about 15,000 yr B.P. This early fauna is comprised of species associated with open ground habitats from within the boreal climatic zone. In other words, the beetles live in the boreal zone, but their specific habitats include meadows, stream banks, and other openings in the forest. The presence of coniferous bark beetles in the assemblage offers evidence for the proximity of coniferous trees. In contrast, Watts reports "clear evidence for tundra vegetation" in pollen spectra from this basal horizon at Longswamp (1979, p. 433).

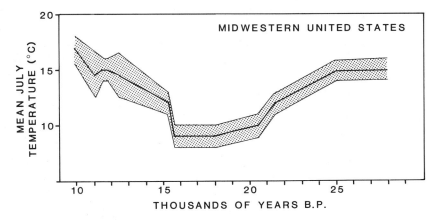

Figure 5.6. Estimate of mean July temperatures during the late Wisconsin and early Holocene, based on insect fossil assemblages from sites in the midwestern United States. (Data from Schwert and Ashworth, 1988.)

A 12,500-yr B.P. fauna was sampled from a kettle pond deposit at Brampton, near Toronto, Ontario (Morgan et al., 1984a). Again, this fauna is characterized by species living in open ground situations within the boreal zone of Canada. Paleobotanical evidence from Brampton is indicative of open ground tundra, including *Dryas integrifolia, Vaccinium uliginosum,* and dwarf birch. Regional plant macrofossil and bark beetle evidence indicates that conifers did not colonize the site until about 500 years later.

This 500-year lag in forest response to climatic warming in the Great Lakes region was also demonstrated at the Gage Street site in Kitchener, Ontario (Schwert et al., 1985). Insect assemblages from 13,000 yr B.P. onward reflected climate characterized by mean summer temperatures greater than 10°C. Today, regions in Canada that experience summer temperatures greater than 10°C support boreal forest, whereas regions with summer temperatures colder than 10°C support arctic tundra (Bryson, 1966). So the insect data suggest that southern Ontario was warm enough to support the growth of conifers from 13,000 yr B.P. onward, but park-like tundra vegetation continued at the site until about 12,500 yr B.P. Ecological succession from spruce forest to pine-dominated forest occurred locally about 2000 years later.

Unlike the European records, the North American insect fossil faunas do not provide any significant evidence for reversals in the late glacial warming trend. The only exceptions to this generalization are chironomid larval faunas from the maritime provinces of Canada (Walker et al., 1991a).

Schwert and Ashworth (1988) and Schwert (1992) have summarized the late glacial insect faunas from the midwestern United States (Fig. 5.6). Fossil beetles from Fort Dodge, Iowa, provide the earliest regional evidence for warming after the Wisconsin glacial maximum, at about 15,300 yr B.P. Additional amelioration through the early Holocene is indicated by midwestern insect faunal assemblages. Schwert and Ashworth (1988) explained the discrepancy between early postglacial environments deduced from insect assemblages and those suggested by paleobotanical reconstructions in the midwestern United States as follows. Pollen and plant macrofossils that accumulated in ice-marginal deposits just after deglaciation reflect tundra-like communities with sedges and *Dryas* and are lacking in trees and shrubs. However, the insect faunas are consistently different; they are analogous to modern faunas from the middle of the boreal forest zone. At each study site, the discrepancy between the plant and insect evidence has been accounted for by invoking a substantial lag in the arrival and establishment of woody plants on deglaciated landscapes. As in central Europe, this lag may represent the time needed for ecological succession from open ground to herbaceous cover, shrub cover, and finally forest cover.

My own work on the late glacial insect faunas of the Rocky Mountain region has been summarized (Elias, 1990b, 1991). The earliest evidence for the start of postglacial warming is a 13,700-yr B.P. assemblage at the Mary Jane site in the Rocky Mountains of north-central Colorado (Short and Elias, 1987). It suggests that rapid warming occurred after 11,500 yr B.P., with summer temperatures becoming at least as warm as those in the present by 10,200 yr B.P. and warmer than those in the present shortly thereafter (Fig. 5.7). Coniferous forest began moving upslope in the Rockies following deglaciation but did not reach its present elevation until 9500–9000 yr B.P. (Short, 1985).

In the Chihuahuan Desert, more than a thousand kilometers south of the southernmost extension of the Wisconsin ice sheets, the insect fossil record shows a clear transition from glacial to postglacial climatic regimes. The chief climatic signal in this desert is the shift from mesic to xeric environments, indicated in insect faunal assemblages starting at about 12,500 yr B.P. (Elias, 1992b). The principal change in plant communities that has been used by paleobotanists as an indicator of the shift to postglacial conditions is the shift of conifers from the lowlands and foothills of the Chihuahuan Desert to higher elevations in regional mountains. This shift is not generally recorded in fossil records older than 11,500–11,000 yr B.P. (Elias and Van Devender, 1992).

Perhaps the most extreme example of plant migration lag documented in North American records is that from southwestern Alaska. Postglacial warming began in this region by 12,500 yr B.P. (Lea et al., 1991). Once again, the insect faunas from this time reflect open ground habitats from within the boreal forest. The insect fossils offer evidence of early postglacial warming to near-

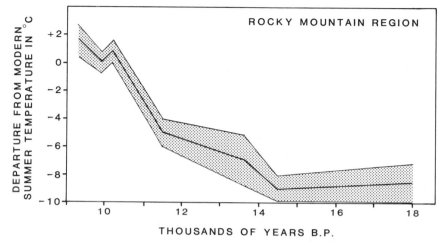

Figure 5.7. Estimate of mean July temperatures during the late glacial period, based on insect fossil assemblages from the Rocky Mountain region. (Data from Elias, 1990b; Elias et al., 1991; Elias, unpublished findings.)

modern levels, indicating summer temperatures warm enough to support the growth of spruce forest (Elias, 1992c). However, spruce forest did not arrive in southwestern Alaska until the mid-Holocene (Short et al., 1992), or about 7000 years after the initiation of postglacial warming.

What kept the forest from advancing into this region? Edwards and Elias (unpublished data) have suggested that effective moisture was insufficient during the first half of the Holocene, as moisture appears to be the limiting factor affecting spruce growth in western Alaska today. Yet the issue of plant migration lag deserves a more detailed explanation.

PLANT MIGRATION LAG

Ever since the fossil insect data first began to offer evidence of rapid response to climate change, the issue of plant migration lag in response to climate change has been the subject of vigorous debate in the literature and at scientific meetings (e.g., Cole, 1985; Markgraf, 1986). The different response times suggested by pollen and other proxy indicators of climate change, such as beetles, have been explained in several ways. Huntley and Webb circumvented the problem of "supposed lags" by proposing that vegetation changes are consistent with climate change "if viewed at the appropriate spatial and temporal scales" (1989, p. 15). On the largest scales (thousands of years and subcontinental regions) vegetation may well be in equilibrium with climate, although the word "equilibrium" seems inappropriate in this context. Such scales may be useful for establishing the broad outlines of change, but the implication seems to be that pollen is inadequate as proxy data for identifying rapid change. Indeed the most recent computations of vegetation response to climate change since the last glacial maximum in eastern North America (Prentice et al., 1992) indicate lags of as much as 1500 years.

The question of vegetation-climate equilibrium has some important implications beyond isolated studies of insect data versus pollen data. One of the most popular trends in paleobotany in recent years has been the application of multiple regression statistical analyses of pollen data as a means of reconstructing paleoclimates, as part of such ambitious projects as CLIMAP (1981) and COHMAP (1988). These reconstructions formally assume that vegetation patterns (albeit on the continental scale) in the late Quaternary (on time scales of 500–1000 years) have been in equilibrium with climate change. However, as Prentice (1986) points out, this assumption appears incompatible with the results of some detailed studies on tree migration. One of these is the investigation of postglacial tree migrations in the Great Lakes region by Davis et al. (1986). They reported evidence that the geographic limit of beech (*Fagus*) was in disequilibrium with climate in southern Wisconsin during the interval 7000–6000 yr B.P., and that hemlock (*Tsuga*) was also in disequilibrium with climate prior to 5000 yr B.P.

Concerning the responses of trees to climate change, Brubaker (1986) noted that population responses of trees to climate are difficult to document, because of their long life span. Brubaker's ecological studies also concluded that tree population changes may lag behind climatic shifts by intervals on the order of decades or centuries.

Bartlein and Prentice (1989) argued that species ranges change slowly in response to synoptic-scale shifts in climate, such as those driven by insolation fluctuations (i.e., orbital forcing: the effect of changes in the earth's orbit on global climate). However, over intervals of about 1000 years or less, dispersal rates, soil processes, and the dynamics of succession are important factors in determining how rapidly a plant taxon can occupy a newly available, climatically suitable region (Prentice, 1986; Ritchie, 1986).

An example of the importance of soil development in vegetation establishment was described by Pennington (1986). The postulated intense warming in late glacial Britain, discussed earlier, preceded the expansion of tree birch by 500–1500 years. Small populations of tree birch were represented by macrofossils early on in the period, so the explanation of the delayed regional expansion appears to lie in differing rates of soil development on a heterogeneous landscape. Precipitation levels may also have been limiting in the early late glacial (Prentice, 1986).

Huntley and Webb (1989) put forward the idea that enhanced seasonality between 12,000 and 6,000 yr B.P., resulting from orbital forcing, can account for the observed behavior of beetles in this time. They believe that, since insects are poikilotherms and inactive in winter, they are limited chiefly by summer temperatures. Hence, their fossils are essentially valid as proxy data only for summer conditions. However, this is not necessarily the case, as MCR studies show (see page 78). Furthermore, it would appear that the length and warmth of the growing season (as measured in growing degree-days) is one of the most important factors controlling the establishment and reproduction, and hence the dispersal capabilities, of both plants and insects in a given region. Perhaps other factors are equally important. Ritchie (1986) has suggested that our perceptions of the nature of vegetation-climate equilibria, both past and present, are based on biased, geographically restricted experience.

Prentice (1983) criticized the evidence for plant migration lag as based on "oversimplified views" of the nature and character of past climate change. A clearer view of biotic response to climate change is undoubtedly needed. When different types of proxy data yield contradictory evidence about past environments, the investigators involved should look on the discrepancy as a learning opportunity. Perhaps more can be grasped about the autecology of the proxy taxa, or we may gain a better understanding of past environments for which there are no modern analogues. Certainly our progress toward these goals is not aided by the dismissal by one group of researchers of the results of others.

6

INSECT ZOOGEOGRAPHY
IN THE QUATERNARY

> The patterns of distributions of animal and plant species [are] the result of a
> complex history that cannot be understood completely without the evidence
> of this fossil record. To ignore the fossil data would be like trying to re-
> construct the plot of a film from a study of its last frame.
> —Russell Coope (1990)

The aim of this chapter is to demonstrate some of the remarkable changes in
insect distributions through the last glacial/interglacial cycle as shown in the
fossil record. By helping investigators to establish rates and patterns of distributional shifts through time, the fossil record is making important contributions to the field of biogeography.

The science of biogeography (including zoogeography, the study of animal distribution patterns, and phytogeography, the study of plant distribution patterns) began in earnest in the nineteenth century. An early student of
the field was Charles Darwin, whose observations on the distributions of
species, past and present, played a key role in the development of his
evolutionary theory. Systematic entomologists began considering insect
zoogeography at about the same time. Notable among these was the American entomologist, John LeConte, who made significant strides in this
developing field as early as 1859.

However, fossil evidence from the Quaternary record had little impact on
the study of insect zoogeography until much more recently. Van Dyke
(1939) offered his considerations on the origin and distribution patterns of
the North American beetle fauna. In his paper, he complained of the lack of
Pleistocene fossil data, except for those from the California asphalt deposits
and Scudder's specimens from Toronto. To his credit, Van Dyke noted that
the available Pleistocene fossils were "usually identical" to the modern
species. Given the lack of fossil data, Van Dyke proceeded to postulate the
origins of modern beetle distributions based solely on evidence gleaned
from their modern ranges.

As Coope (1990) noted in the quotation at the start of this chapter, it seems intuitively obvious that it may be impossible to discern the history of species movements through time based only on where they have settled down most recently. Nevertheless, numerous zoogeographers continue to attempt to reconstruct insect distributional histories and to erect biogeographic and evolutionary theories without consideration for (or perhaps in spite of) the fossil evidence. For instance, biogeographers have suggested that the evolution of populations in refugia may be an important mechanism for speciation and for the production of geographic distribution patterns (Noonan, 1985). Noonan (1988) suggested that Pleistocene glaciations may have had the general effect of causing extinction of many of the more sedentary insects of North America, and that Pleistocene environmental stresses on North American insects probably selected for relatively vagile taxa. Yet, as we have seen, there is little or no fossil evidence of insect extinctions caused by Pleistocene glaciations, even in the more sedentary insect groups (Coope, 1978).

Likewise, Crowson (1981) suggested that many flightless beetles have very limited powers of dispersal, to the extent that they have practically no power to extend their range across even narrow belts of unfavorable terrain. This may be true for some taxa, but the fossil record suggests that many flightless species have shifted distributions dramatically through Quaternary time.

As Morgan and Morgan (1980) pointed out, with the accumulation of fossil records it is now becoming possible to raise knowledge of faunal histories and speciation above the realm of speculation. Let me issue one caveat before we begin. Although the fossil record broadens our knowledge considerably, the records of most individual species cover only a small proportion of those species' life span. The fossil record of a given beetle species may cover 100,000 years, but this would represent only 5% of the total life span of the species, which may have been two million years. Van Dyke (1939) anticipated this problem when he stated

Most of our orders of insects were well established in Permian times, at the end of the Paleozoic. Thus, the insects being of infinitely older stock than the higher vertebrates, we are justified in concluding that the history of their earlier migrations and distribution goes back very much further in geological times than does that of the mammals and birds.

An important constraint in postulation of zoogeographic events is the dispersal power of insects. This factor was touched upon in Chapter 5 as it related to insect response to climate change. In a general summary of dispersal mechanisms in beetles, Crowson (1981) recognized several categories of beetles in respect to their powers of dispersal, ranging from the very vagile to the extremely sedentary. As indicated later in this chapter, even some supposedly sedentary groups have moved about considerably during the late Quaternary.

PLEISTOCENE RECORDS FROM EUROPE

The British fossil record provides the best-documented history of distributional shifts in response to climate change. Fossil assemblages from the last (Ipswichian) interglacial and interstadials within the last glaciation provide the basis for inferring invasion of Britain by species that today have a Mediterranean range (Coope, 1990). These beetles have a variety of ecological requirements, and some have such restricted modern distributions that they have been labeled endemic Mediterranean species.

Ground beetles such as *Bembidion grisvardi* and/or *B. ibericum* (modern range shown in Fig. 6.1) live in open habitats with sparse vegetation. Their presence in Britain at the end of the last glaciation (circa 13,000 yr B.P.) is indicative of sudden climatic warming and an open, poorly vegetated landscape

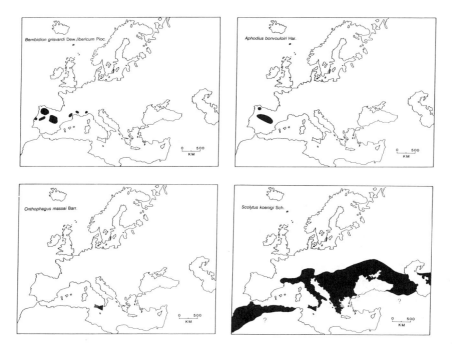

Figure 6.1. Modern distributions of thermophilous species of beetles found in British interstadial and interglacial deposits. (After Coope, 1986.)

with poor soil development. This environment scarcely exists in modern Britain, but must have been much more widespread at the end of the last glaciation (Coope, 1990). *B. grisvardi* was also identified from a late glacial assemblage at Champreveyres, Switzerland (Coope and Elias, unpublished data).

Another thermophilous ground beetle found in last interglacial deposits in Europe is *Oodes gracilis*. This species lives in aquatic environments, particularly reed swamps, where adults crawl on submerged plant stems. Interglacial-age fossils of *O. gracilis* have been found in Britain (Coope, 1990) and in Byelorussia (V. I. Nazarov, written communication, 1991). In both instances, the presence of *O. gracilis,* in combination with other thermophiles, indicates climatic conditions warmer than those in the present.

During past episodes of warm climate in Britain, the dung beetle fauna was enriched by a number of thermophilous species. Among these are the Mediterranean beetles *Aphodius bonvouloiri* and *Onthophagus massai* (Fig. 6.1). *A. bonvouloiri* was one of the most abundant dung beetles in warm interstadial assemblages in Britain (Girling, 1974). At the termination of the Upton Warren Interstadial, *A. bonvouloiri* was replaced in the British fauna by *A. holdereri,* the dung beetle discussed in Chapter 1 that today is found only on the Tibetan Plateau (Coope, 1973). Perhaps the most exotic thermophilous dung beetle in the British Pleistocene faunas is *O. massai,* a species found today only on the island of Sicily. Here modern distribution might suggest endemism, but the fossil record refutes that hypothesis. As Coope (1990) pointed out, such an isolated modern geographic range could just as well be the last stand of the species as its place of birth.

Another scarab beetle found in last interglacial deposits in Britain is the genus *Drepanocerus*. This group is known today only from Africa south of the Sahara Desert, the Indian subcontinent, and southeast Asia (Coope, 1979). Its modern range is a fragmented remnant of a once wide distribution that was apparently disrupted during the last glaciation.

Even host-specific phytophages can provide useful zoogeographic clues to past environments. The bark beetle *Scolytus koenigi* lives under the bark of maple (*Acer* spp.) and allied trees, and is found today across southern Europe and northwestern Africa (Fig. 6.1). It was found in British fossil assemblages from the last interglacial, as were pollen and macrofossils of the host plants (Coope, 1990).

Cold-adapted faunas replaced these thermophilous species during glacial cycles in Britain. Beetles with arctic or arctic and alpine affinities were the most abundant elements in the glacial assemblages. *Diacheila polita* is an arctic ground beetle found in British fossil records as well as in glacial assemblages from eastern Europe and North America south of the last glacial ice

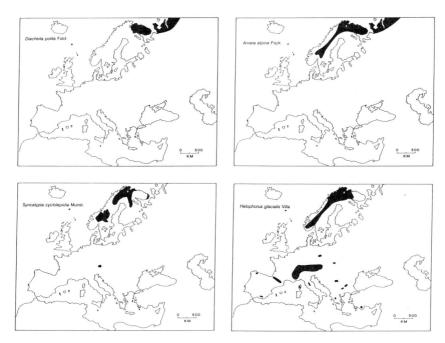

Figure 6.2. Modern European distributions of cold-adapted species of beetles found in British glacial deposits. (Data from Lindroth, 1960.)

sheets. In Eurasia today, *D. polita* is found only in arctic regions from the Kola Peninsula eastward (Fig. 6.2). In North America, it is found today only in arctic and subarctic regions of Alaska and the Yukon Territory, with isolated populations on alpine tundra in the Alaska Range.

Another ground beetle, *Amara alpina,* is an arctic and alpine species today, with populations extending south along mountains in Scandinavia (Fig. 6.2). It is the northernmost ground beetle in the modern North American fauna (Fig. 6.4), but isolated populations are found on mountaintops as far south as northern New Mexico in the Rocky Mountain chain and on top of the few high mountains in the northern Appalachians. It has been found in numerous glacial-age deposits in Britain (Coope, 1977), as well as in late Weichselian assemblages from southern Sweden (Lemdahl, 1988c, 1991a).

The pill beetle, *Syncalypta cyclolepidia,* was a widespread inhabitant of Europe during the last glaciation. It has been found in British fossil faunas as well as in glacial-age assemblages dated to near the end of the last glaciation at Lobsigensee, Switzerland (Elias and Wilkinson, 1983). Today this moss-feeding beetle occupies arctic and alpine regions of Scandinavia, and has isolated populations in the Alps of Austria (Fig. 6.2).

Another cold-adapted beetle that was apparently widespread in Europe during the last glaciation is the water scavenger beetle, *Helophorus glacialis.* The modern distribution of this species in Europe reflects its retreat to cold climate regions, both in the north and on mountaintops throughout the rest of Europe (Fig. 6.2). In the last glaciation, the fossil record indicates that it lived in Sweden (Lemdahl, 1991a), Britain (Coope, 1977), France (Ponel and Coope, 1990), and Switzerland (Elias and Wilkinson, 1983; Coope and Elias, unpublished data).

The relative severity of various cold stages has been clarified by the presence of certain beetle species in the British fossil record. For instance, the water scavenger beetle, *Ochthebius kaninensis,* has been found in deposits from a cold period just prior to the Upton Warren Interstadial. Today, it is known only from the Kanin Peninsula in the Russian Arctic (Coope, 1990).

The carabid species in the subgenus *Cryobius* (genus *Pterostichus*) are cold adapted and occupied Europe during the last glaciation. This group is no longer found anywhere in western Europe. Most species live in eastern Siberia and arctic North America, with many species known today only from Alaska (Ball, 1966). Coope (1990) points out that the concentration of *Cryobius* species in a given area (such as Siberia and Alaska) today may be more indicative of their common ecological requirements than of an original center of dispersal. *Cryobius* fossils have been found in glacial deposits from Britain (Coope, 1977) and from mid-Weichselian assemblages at the Niederwenigen site in Switzerland (Elias and Schlüchter, unpublished data). Nazarov (written communication, 1991) reported *Cryobius* fossils from middle Pleistocene (circa 440,000-yr B.P.) deposits in Byelorussia, in association with *Diacheila polita* and other indicators of arctic conditions. These European examples show that it may be unwise to describe any beetle species as endemic to a given region until its fossil history has been determined.

Dramatic distributional shifts have also been documented elsewhere in the Pleistocene beetle fauna of Eurasia. For instance, Kiselyov (1973) described distributional changes in late Pleistocene beetle faunas in the Ural Mountains. His work showed that the Transuralian region was once home to a unique mixture of xeric-adapted species found today in steppe regions of Kazakhstan and arctic/subarctic species found today in eastern Siberia and Mongolia.

Likewise, Pleistocene insect assemblages from eastern Siberia contain mixtures of steppe and arctic tundra elements (Kiselyov and Nazarov, 1984). These species have undergone continental-scale shifts in distribution within the last glacial/interglacial cycle. For instance, the weevils in deposits from the Kolyma lowland include the genera *Coniocleonus* and *Stephanocleonus,* confined today to steppe and montane-steppe regions well to the south in Siberia and Mongolia. Some steppe-associated weevils still live in isolated habitats on south-facing slopes in Yakutia, the Taimyr Peninsula, western Chukotka, and Wrangell Island (Korotyayev, 1977).

The Beringian steppe-tundra and its insect fauna are treated more fully in Chapters 9 and 11. The arctic species found in many of these assemblages include *Amara alpina* and several species of *Cryobius,* although species of the latter group occur only in intervals of more mesic climate.

NORTH AMERICAN STUDIES

Fossil evidence indicates that North American insects underwent large-scale distributional shifts during the Quaternary. For instance, the last glaciation forced the dispersal of cold-adapted insects into two major refugia (Morgan et al., 1984a). Unlike in Europe, not all high-latitude regions in North America were glaciated during the Pleistocene. The lowlands of Alaska, the Yukon Territory, and the westernmost edge of the Canadian Northwest Territories were free of ice for much of the Quaternary. Moreover, as sea levels lowered during glaciations, the shallow continental shelf regions between Alaska and Siberia became dry land, connecting the ice-free regions on either side of the Bering Strait. This region served as a refugium for arctic biota. Eric Hultén put forward the idea of a Beringian refuge in 1937, based solely on the modern distribution patterns of arctic plants.

A second refugium for cold-adapted biota was located south of the ice sheets and extended from the state of Washington through Montana and the Dakotas into southeastern Wisconsin, Illinois, northeastern Pennsylvania, and New York. These two refugia have been well established in the fossil record. A third refugium on the exposed continental shelf regions off the east coast of North America has been postulated, although as yet no insect fossils have been recovered (Morgan et al., 1984a) to support the existence of this refugium. However, vertebrate fossils have been dredged up from the continental shelf.

Dispersal of Beetles from the Southern Refugium

The insect faunas of late Wisconsin age from the southern refugium are composed of species with a variety of modern distributions. No arctic-style climate occurs today in the refugial region, except on mountaintops in the Rocky Mountains and isolated peaks in the northern Appalachians. Accordingly, the cold-adapted fauna that inhabited the refugium during the Wisconsin

Figure 6.3. "*Diacheila polita* in Iowa, 12,000 yr B.P." (After Faukse, Ashworth, and Schwert, unpublished. Courtesy Donald P. Schwert.)

glaciation was extirpated from lowlands south of the ice sheets when climate warmed at the end of the last glaciation. A few cold-adapted beetle species became established on alpine tundra regions in the aforementioned mountain regions, following local deglaciation of mountain ice caps. One of these is the ground beetle species *Amara alpina*. In contrast *A. alpina* did not become established in the European Alps at the end of the last glaciation. It is adapted to cold, xeric conditions, such as are found in the high arctic. Similar conditions persist on some high mountains in North America, but perhaps the Alps are too moist on the whole to support *A. alpina* today.

Most other cold-adapted beetle species were extirpated from the southern refugial regions (Schwert and Ashworth, 1988). The retreat of the Wisconsin ice sheets was slow, and the ice-proximal regions became too warm to support these species (Fig. 6.3). In some regions, boreal forest encroached on the retreating ice margins. Following deglaciation, Beringian populations of cold-adapted beetles have recolonized the central and eastern arctic regions of North America.

Not all beetle groups have been equally successful in becoming established in previously glaciated landscapes. Predaceous beetles (especially members of the Carabidae and Staphylinidae) were relatively effective in recolonizing

Canada. One reason for this is that they exhibit good dispersal abilities; hence they were able to reinvade newly exposed regions rapidly, following deglaciation (Campbell, 1980).

One of these invaders is the ground beetle species *Asaphidion yukonense,* known today from central Alaska south through the Yukon Territory and northern British Columbia, with isolated populations in the Rocky Mountains of Alberta (Fig. 6.4). *A. yukonense* lives in open areas with scarce vegetation consisting of small moss patches. It is easy to envision this type of habitat in recently deglaciated landscapes, and in fact fossils of this species have been found in this depositional context from sites in midwestern and eastern North America (Morgan et al., 1984a).

The species richness of the modern arctic beetle fauna declines markedly from west to east. This situation appears to be due to migrational lag following deglaciation and probably not because of differences in productivity in the respective landscapes (Schwert, 1992). The beetle fauna of the eastern arctic has yet to achieve its potential diversity. One of the factors curtailing the easterly dispersal of tundra insects from Beringia has been northward expansion of trees in the Mackenzie River Valley during the Holocene. However, strong westerly winds prevalent across the northern tundra regions should have facilitated eastward spread of insects through aerial dispersal, despite the presence of the Mackenzie forest barrier (Danks, 1981).

Hudson Bay has also apparently been an effective barrier in eastward dispersal of beetles in the Holocene. The ground beetle species *Pterostichus (Cryobius) caribou* is one of many arctic beetle species that exhibit a modern distribution including northern Alaska and northern Canada, with an abrupt termination on the western shores of Hudson Bay (Fig. 6.3). Another factor limiting the postglacial reinvasion of the eastern arctic is that the central part of the Labrador-Ungava peninsula was not fully deglaciated until after 7000 yr B.P. (Dyke and Prest, 1986), whereas much of the western arctic was deglaciated by 10,000 yr B.P.

In contrast to arctic species, the boreal insect fauna that survived the last glaciation south of the ice sheets was probably able to disperse northward along the margin of receding ice. An example of this type of dispersal is provided by the boreal ground beetle species *Notiophilus borealis* (Fig. 6.3). The northward shift of the boreal fauna has been documented in several midwestern fossil assemblages (Morgan, 1987; Schwert and Ashworth, 1988).

The modern distributions of some of the beetles that have recolonized arctic North America suggest modes of dispersal that are quite intriguing. The rove beetle species *Holoboreaphilus nordenskioeldi* is known today from arctic Canada and Alaska, including the northern tip of the Labrador-Ungava penin-

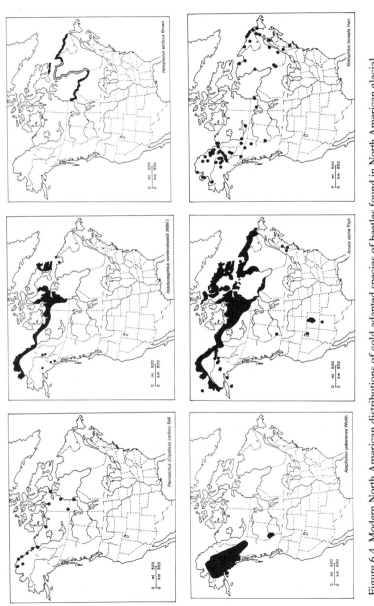

Figure 6.4. Modern North American distributions of cold-adapted species of beetles found in North American glacial deposits. (Data for *P. caribou* from Ball, 1966; map of *H. nordenskioeldi* after Morgan et al., 1984b; map of *H. arcticus* after Morgan, 1989; map of *A. yukonense* after Morgan et al., 1984a; data for *A. alpina* from Lindroth, 1968; data for *N. borealis* from Lindroth, 1961.)

sula and southeastern Baffin Island (Fig. 6.3). It has been found in only two Wisconsin-age deposits from south of the glacial margins. One of these is an ice-proximal deposit from Titusville, Pennsylvania (Totten, 1971), dated roughly between 41,000 and 35,000 yr B.P. The other occurrence is from an 11,400-yr B.P. assemblage at Marias Pass, Montana (Elias, 1988a). Its modern presence on Baffin Island and northern Ungava is puzzling. Morgan et al. (1984b) proposed three hypotheses to explain its dispersal to the eastern Canadian arctic. The first is easterly postglacial dispersal from population centers in Alaska. The second is dispersal on drifting pack ice from the last vestiges of disintegrating Laurentide ice in eastern Keewatin, across Hudson Bay to Baffin Island. The third is dispersal by drifting on pack ice from the northern tip of Labrador, following local deglaciation (circa 8500 yr B.P.). Perhaps additional fossil studies on the Labrador coast will help elucidate the postglacial movements of *H. nordenskioeldi*.

Fossils of the water scavenger beetle species *Helophorus arcticus* are known from a number of Wisconsin-age fossil sites in central and eastern North America, south of the Laurentide ice sheet. It is found in fossil and modern assemblages generally indicative of climatic conditions associated with northern treeline (i.e., mean July temperatures at or near 10°C). Morgan (1989) postulated the movements of *H. arcticus* from the Sangamon interglacial onward. With the onset of the Wisconsin glaciation, northern Canadian populations were forced south in eastern North America. Lowered sea level probably provided substantial habitats along the exposed continental shelf from Labrador south to New England. Inland, *H. arcticus* inhabited the shores of proglacial lakes, a habitat not currently in existence, but one that most certainly has been available throughout much of the Quaternary history of North America. (In this context, it is important to keep in mind that glaciations account for the majority of Quaternary time and that interglacials are relatively short, on the order of 10,000 years.) With the onset of deglaciation, *H. arcticus* migrated north, following the shores of huge proglacial lakes formed by meltwater from the receding Laurentide ice sheet. This dispersal route allowed *H. arcticus* to migrate eastward in the Great Lakes region, onto the southern shores of the late glacial Champlain Sea in Quebec and north from there as the ice continued to recede into Labrador. Other than *H. arcticus,* the only cold-adapted beetle to have succeeded in postglacial dispersal into northern Quebec along eastern coastal tundra is *Amara alpina* (Schwert, 1992). *H. arcticus* has also turned up in mid-Pleistocene deposits from England (G. R. Coope, personal communication, 1993).

Dispersal of Rocky Mountain Insects

The Rocky Mountains have been a refuge for many cold-adapted species of both plants and animals. Unlike the fauna in midwestern and eastern North

America, where most cold stenotherms were extirpated after the Wisconsin glaciation, some of the Wisconsin fauna that lived in close proximity to the Rocky Mountains was able to track suitable habitats upslope to the alpine tundra zone. Hence, several cold-adapted species that lived on the plains south of Denver toward the end of the last glaciation have retreated to high elevations in the Colorado Front Range (Elias, 1986). Nevertheless, numerous extirpations of cold-adapted species have occurred in the Rocky Mountain region.

Thirteen species of beetles that are no longer living in the Rocky Mountain region have been identified from regional fossil assemblages dating to the interval from 18,000 to 2500 yr B.P. Although additional modern collecting in the region may reveal extant populations of some of these beetles, it appears likely that most of them are indeed gone. Six of the eight species eliminated before the Holocene have shifted their distributions to the north. These are all cold-adapted animals, found today either in the arctic or in boreo-arctic regions of Canada and Alaska.

Of the species extirpated from the Rocky Mountain region during the Holocene, only one has shifted its distribution to the north. The riparian ground beetle species *Bembidion rusticum* lives now only in British Columbia and northward, more than 2600 km from the fossil site at Lake Emma, Colorado (Elias et al., 1991). Two other species in this category have modern ranges in eastern North America, one in the Appalachians and one in the grasslands of the Midwest. These distributional shifts appear not to be due to postglacial warming, although some extirpation occurred during the interval of Altithermal warming in the Holocene. They may be related to changes in moisture.

The trends observed from the Rocky Mountain fossil record also hold true for fossil insect faunas from the Chihuahuan Desert. During the late glacial interval, most regional extirpations were of species that live to the north of that desert today, whereas in the Holocene most of the regional extirpations were of species that live either to the east or to the west of the Chihuahuan Desert.

Many temperate species of beetles were able to take advantage of cool, moist habitats established in the Chihuahuan Desert during the last glaciation, the spine of the Rocky Mountains serving as a corridor to the desert regions.

In his biogeographic analysis of the present-day distribution of the carabid genus *Harpalus,* Noonan (1990) noted that the species in western North America occur in lowland habitats in the north and shift progressively upward into mountains to the south, as desert and other xeric environments replace mesic lowland habitats. This ability to exist in montane habitats facilitates broad distribution by enabling species to expand their ranges southward. *Harpalus* species have persisted in the southern Rocky Mountains at least in part because of their ability to track suitable habitats up- and downslope as climates changed.

Postglacial Shifts in the Beetle Fauna of the American Southwest

The study of insect fossils from packrat middens preserved in the arid climates of southwestern deserts is beginning to bring to light the insect faunal history of this region. Although fossil insect research is now underway for sites in the Great Basin (Elias et al., 1992b), the most intensively studied region thus far, with 191 midden insect fossil samples from 27 sites, is the Chihuahuan Desert (Elias and Van Devender, 1990, 1992; Elias, 1992b). Late Pleistocene faunas comprised mixtures of temperate and desert species not seen in any one region today. What has become of the late Pleistocene beetle fauna of this region? Some species now live elsewhere in the Chihuahuan Desert. Others live in other regions. In fact, an examination of the modern distributions of several species permits the postulation of movement in every possible direction: several species now live to the west, into the Sonoran, Mohave, and Great Basin deserts; some now live to the east, on the plains of Texas and Oklahoma; others now live to the north and northeast, in the prairie regions on the east flank of the Rockies; a few species now live to the south, in Mexico.

Sufficiently detailed regional analyses of late Quaternary Chihuahuan Desert insect faunas are available to allow inferences as to the timing of their distributional changes (Elias, 1992b). These shifts follow no single pattern (Fig. 6.5). Some species, such as the dung beetle *Onthophagus lecontei,* may have begun migrating as early as 15,000 yr B.P. It was last recorded from sites in the southern Chihuahuan Desert at 12,500 yr B.P., and its range is now restricted to central Mexico. Another dung beetle, *Onthophagus cochisus,* appears to have begun moving northward from the Big Bend region at 19,800 yr B.P. It was last recorded from the Trans-Pecos region at 11,000 yr B.P. and is now found only in the higher elevations (1500–2400 m) of the Chiricahua Mountains of southeastern Arizona and in the mountains of northern Sonora, Mexico (Howden and Cartwright, 1963). The ground beetle species *Amara chalcea* apparently departed the southern Chihuahuan Desert by 12,500 yr B.P. but remained in the central part of the desert until 9400 yr B.P., with at least one population surviving in the hills of Last Chance Canyon, New Mexico, until 1500 yr B.P. Today this species is found only in the grasslands of the Great Plains region, on dry, open, sandy ground (Lindroth, 1968).

Some of the most puzzling distributional changes of Pleistocene Chihuahuan Desert beetle species are those of the cave-dwelling ground beetles in the genus *Rhadine*. These beetles are dorsoventrally flattened, a characteristic that enables them to maneuver through narrow cracks in caves. They are flightless and depigmented; most have reduced eyes. Pleistocene fossils have been identified from assemblages in the Chihuahuan Desert (Elias and Van Devender, 1990, 1992), but modern records of some of these species are limited to caves in

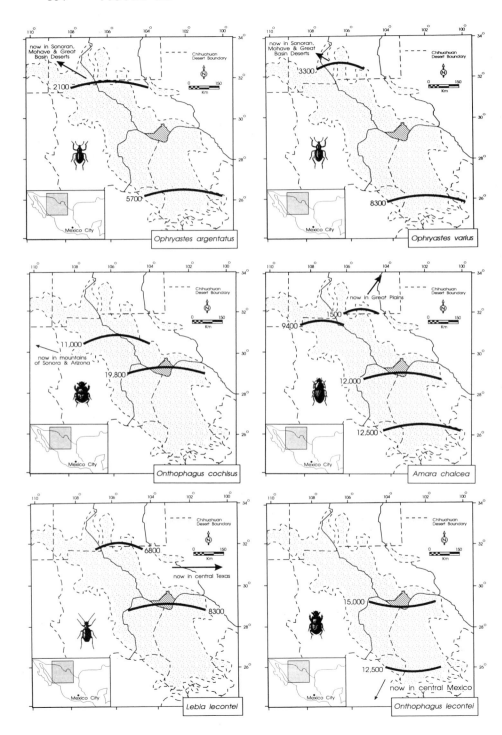

central Texas and Oklahoma. Needless to say, there are no known subterranean connections between these far-distant cave systems. Another puzzling shift involves *Rhadine longicolle,* known today only from Carlsbad Caverns, New Mexico, and nearby caves within Carlsbad Caverns National Park (Barr, 1960). Based on the modern evidence alone, one might suppose that *R. longicolle* has always been restricted to the caves of the Carlsbad region, but fossil specimens were identified from the Big Bend region of Texas, about 400 km to the southeast (Elias and Van Devender, 1990). Barr (1960) discussed the possibility of geographic isolation of ancestral stocks in west Texas during the late Pliocene or early Pleistocene, but the fossil evidence indicates that these beetles are able to migrate substantial distances within at most a few thousand years.

The fossil insect record of the Chihuahuan Desert indicates that sedentary, flightless beetles (such as the heavy-bodied weevils in the genus *Ophryastes*) have undergone marked distributional shifts in the American southwest within the space of a few centuries. Moreover, even highly specialized cave dwellers such as the ground beetles of the *Rhadine* group have somehow managed to move from one cave system to another in response to changes in late Quaternary environments. It appears certain that packrats are not agents of dispersal of beetle remains from one cave to another. Insects are not an important part of packrat diets (Armstrong, 1982). Moreover, modern studies with packrats have shown that beetle fragments found in middens, including those representing cave-dwelling species, have not been consumed by the packrats (Elias, 1990a). Cave-dwelling insects probably become trapped in the sticky mass of packrat feces and urine as they forage near the mouth of the cave where the packrat makes its nest.

Although the fossil record has shed new light on some important questions regarding insect Quaternary zoogeography, it also has generated a new batch that are at least as exciting as the previous ones. It is hoped that the results reviewed here will stimulate zoogeographers and paleoentomologists to cooperate more fully in trying to answer these. The tangible data on past distributions that are made available in the Quaternary fossil record may disprove some zoogeographic theories, but they also take us many steps forward in our understanding of the subject.

Figure 6.5. The timing of distribution shifts within and extirpation from the Chihuahuan Desert region for six beetle species identified from late Quaternary packrat middens. Lines curving to the north show proposed southernmost boundaries of species populations in the Chihuahuan Desert at the times indicated (in thousands of radiocarbon years before present). Lines curving to the south show proposed northernmost boundaries. (After Elias, 1992a.)

7

THE USE OF INSECT FOSSILS IN ARCHAEOLOGY

> Entomological evidence has seldom been collected in archaeological investigations, although its usefulness seems promising.
> —Samuel Graham (1965)

Well-preserved insect fossils have been found in many archaeological settings. Along with other types of biological proxy data (e.g., plant macrofossils, vertebrate remains, pollen), fossil insects are now making an important contribution to the reconstruction of both the natural and the anthropogenic environments associated with archaeological sites, supplying evidence about human life-styles and living conditions. Indeed, the biological evidence sometimes reveals far more information than can be obtained from artifacts.

As with paleontological studies, the "home" of research on insects from archaeological sites is the British Isles. I will discuss the British record in more detail shortly. For now, I discuss the types of anthropogenic deposits that have yielded good insect results.

TYPES OF ANTHROPOGENIC DEPOSITS YIELDING INSECTS

Among the most productive anthropogenic deposits for the paleoentomologist are waterlogged sediments at sites of human occupation, such as are common in Britain and elsewhere in northern Europe. Insect exoskeletons in waterlogged sediments show excellent preservation and are sufficiently abundant at some sites to nearly overwhelm the investigator. For example, a medieval moat surrounding a manor house at Cowick, Great Britain (Fig. 7.2, No. 15) yielded 224 taxa of beetles and caddisflies (Girling and Robinson, 1989). The insect fauna from the trackway peats at Thorne Moor (Fig. 7.2, No. 51) comprised 340 taxa in 73 families (Buckland, 1979). As discussed in Chapter 2, insects, as well as other organic detritus, are readily preserved in situations in which normal decomposition is retarded owing to lack of oxygen. Numerous insect studies have been based on these types of sediments, either from archaeological sites themselves or from localities adjacent to (and of the same age as) such sites.

Examples of saturated organic deposits yielding insect fossils include anthropogenic features, such as ancient trackways through bogs and other wet lowlands, organic detritus from wells, and ancient occupation horizons below the current water table. Some nonanthropogenic organic repositories, such as ponds, lakes, and peat bogs, may be situated in close proximity to archaeological sites and may contain insect faunas that are contemporaneous with nearby human occupation.

Domestic debris from under floors and in trash middens, latrines, sewers, barrow pits, and so on has also yielded abundant insect remains. The garbage and sewage of past generations offer many clues to their life-styles, sanitation, animal husbandry, and land use (Moore, 1981; Osborne, 1983).

Preservation of insects is also excellent in both cold (arctic/subarctic) environments and hot, dry environments, such as desert caves, once again, as a function of retarded decomposition in these environments.

Insect fossils have been recovered from a variety of archaeological contexts. Insect remains have been sampled from mummies from Greenland (Bresciani et al., 1983, 1989), Egypt (Alfieri, 1931; Curry, 1979; David, 1984), and the Aleutian Islands of Alaska (Horne, 1979). Insect fossils have been removed from the remains of "bog people," some of whom were victims of ritual murders whose bodies were thrown into bogs. The exoskeletons of stored product pest insects have been recovered from Pueblo Indian cliff dwellings (Graham, 1965), the tombs of Egyptian pharaohs (and the tombs of less regal Egyptians) (Solomon, 1979), and numerous medieval dwellings in Europe.

ENVIRONMENTAL INDICATIONS FROM INSECTS

The ecological sensitivity of insects makes them useful indicators of past environments, both natural and anthropogenic. Evidence from Pleistocene assemblages paved the way for archaeoentomological studies. In Europe, the majority of Holocene paleoclimate interpretations based on insects as well as other biological proxy data are confounded by anthropogenic effects on landscapes, especially forest clearing (Osborne, 1976). Buckland (1979) has asserted that the transformation of most of Europe from a wholly forested landscape to one of artificially maintained meadows, farm fields, and urban

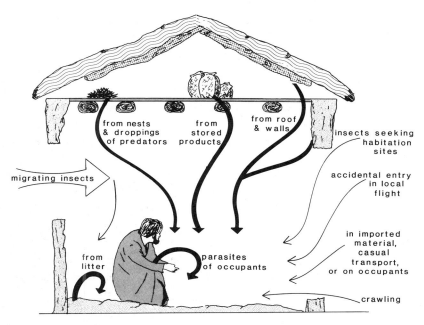

Figure 7.1. Potential sources of insect remains in dwellings. Broad arrows represent insects originating within the building; slim arrows represent insects originating from outside. (After Kenward, 1985b.)

environments represents the most dramatic nonclimatic environmental change of the Quaternary. Hence most British archaeological sites contain a mixture of synanthropic and nonsynanthropic biota, and it is often difficult to make clear separations between these two components (Kenward, 1975, 1976, 1978, 1982). Kenward (1985b) studied indoor and outdoor insect taxa found in modern buildings and so helped to define potential sources of insects deposited in ancient dwellings (Fig. 7.1).

Insect remains can provide a wealth of archaeological detail and may yield information on human activities and environmental conditions not brought out by other types of data. As already noted, stored product pests, as well as insect scavengers and phytophages, provide unique documentation of human foods, food storage and use patterns, food types, building methods, and other details of domestic life, farming, and animal husbandry. Human parasites and synanthropes offer evidence of sanitation conditions, health, crowding, and human dispersal patterns. Carrion beetles, flies, and other insects provide forensic data on human and animal corpses. I will develop each of these themes on a regional basis later in this chapter.

SPECIAL METHODS FOR ARCHAEOLOGICAL SITES

Buckland (1976a) provides a useful first account of methods for the sampling of insects in archaeological contexts. More recent treatments are found in Coope (1986) and in Buckland and Coope (1991).

In general, sediments containing insect fossil samples should be greater than 5 cm thick to provide adequate numbers of specimens per sampling horizon. However, this is not a practical requirement in all situations. For instance, some cultural horizons are only a few millimeters thick. Several kilograms of sediment at one site may yield few samples, whereas relatively small samples at another site may yield an enormous number. Osborne (1969) extracted a huge insect fauna from a small sample from the Wilsford Shaft in southern Britain (Fig. 7.2, No. 56), but I have obtained only twenty identifiable insects after screening several thousand kilograms of sediment from the Lubbock Lake site in Texas (Fig. 7.6, No. 15). There are no hard and fast rules for sampling intervals; the paleoentomologist and archaeologist should work together to develop appropriate sampling strategies.

One must pay careful attention to disconformities, whether natural or anthropogenic. Even a single event (such as the digging of a posthole) may confound a sedimentary sequence beyond recognition. Whenever possible, samples should be taken from newly exposed horizons to avoid modern contamination. Needless to say, it is best if the fossil insect worker does the sampling. If this is impossible, then the archaeologist should take as much material as possible and seal it in plastic bags for delivery to the paleontologist (Buckland, 1976a).

Some problems are inherent in archaeoentomology. Perhaps chief among these is the lack of sampling for insects by archaeologists, which may be due to ignorance ("you mean there might be beetles in my sediment samples?") or neglect. Buckland (1976a) noted that, despite the fact that insect fragments are clearly visible during many excavations, they have often been neglected by workers who are oriented toward artifacts rather than environments. Another problem is prejudice on the part of archaeologists or stubborn adherence to older methods ("a pollen diagram is all we ever need" or "that's the only thing Professor Glockenspiel ever sampled for . . . his methods are good enough for me!"). In many instances, there is simply a lack of time or (more frequently) a lack of money for interdisciplinary studies including insects. In these days of rescue archaeology, the problem is particularly troublesome.

A third serious problem is the lack of suitable publication of the final results of the handful of insect studies that are performed. Many archaeological journals will not accept manuscripts on insect fossil studies. Sometimes reports made by paleoentomologists to archaeologists are too brief to be suitable for publication. In many cooperative studies, the entomologist is not in charge of his or her data set: the archaeologist is given the responsibility of publishing the results of a multidisciplinary study.

BRITISH STUDIES

British insect fossils have been studied from archaeological sites ranging in age from 10,000 yr B.P. through the nineteenth century. More than 100 sites have thus far been investigated for insects. Table 7.1 summarizes the publications of these studies that are more or less readily available to the interested reader.

Only one Upper Paleolithic site has been discovered in which insects are preserved in an archaeological context. This is the Messingham site (Fig. 7.2, No. 37) in north Lincolnshire (Buckland, 1984). Coversand deposits (sand layers laid down by wind storms during a glacial stadial) yielded stone artifacts in conjunction with a nonsynanthropic fauna indicative of a harsh, treeless landscape at circa 10,280 yr B.P.

An extensive temporal gap in the fossil insect record extends from the Upper Paleolithic to the early Neolithic period. At Hampstead Heath, London (Fig. 7.2, No. 28), the insect fauna registered a change from natural forest to cleared sites used for cultivation and grazing (Girling and Greig, 1977, 1985). The insects from this assemblage include lime feeders, ivy feeders, holly feeders, and oak leaf miners. The Neolithic elm decline was noted at the site, as was the presence of the elm bark beetle, *Scolytus scolytus* (Girling and Greig, 1985).

One of the best-documented insect faunal successions of the Bronze Age was discovered at Thorne Moor in South Yorkshire (Fig. 7.2, No. 51) (Buckland and Kenward, 1973; Buckland, 1979). A trackway, dated at 3090 yr B.P., was discovered in a peat bog. This date corresponds to the time of regional forest clearing and the rise of cereal pollen in the regional vegetation. The trackway peat contained beetles from open water and bogs, a finding that suggests that local flooding may have killed the forest. As with other trackway sites, waterlogging of the landscape eventually overwhelmed local farming; it may have been the result of hydrologic changes associated with the rivers of the Humber Basin. Beetle species that are now extinct or extremely rare in Britain provide evidence of old, mature forest. This type of primeval forest and its associated insect fauna do not exist in Britain today and are also rare elsewhere in Europe.

The history of the systematic destruction of primeval forest in Britain is being reconstructed using several methods. Apparently, nearly all primeval forest had been cleared from the English Midlands by the end of the Roman period. Secondary forests, needed for firewood, were preserved for hunting

Table 7.1

British archaeological sites with insect fossil analyses.

Site	Reference(s)
1. Alcester	Osborne (1971a)
2. Alchester	Giorgi and Robinson (1985), Robinson (1975)
3. *Amsterdam* shipwreck, Folkestone	Hakbijl (1987)
4. Appleford	Robinson (1981a)
5. Baginton	Osborne (1975)
6. Barnsley Park	Coope and Osborne (1968)
7. Bearsden	Dickson et al. (1979)
8. Beverley	Hall and Kenward (1990)
9. Birmingham	Osborne (1981a)
10. Breidden Hill	Musson et al. (1977)
11. Carlisle	Kenward (1984)
12. Chichester	Girling (1989)
13. Church Stretton	Osborne (1972)
14. Cirencester	Osborne (1982)
15. Cowick	Girling and Robinson (1989)
16. Crossnacreevy, County Down	Girling (1973–1974, 1974)
17. Denny Abbey	Robinson (1980)
18. Doncaster	Smith (1978)
19. Droitwich	Osborne (1977)
20. Dublin	Coope (1981), O'Connor (1979, 1987)
21. Durham	Kenward (1979a)
22. Empingham	Buckland (1981b)
23. Farmoor	Robinson (1979a)
24. Fishbourne	Osborne (1971b)
25. Fisherwick	Osborne (1979)
26. Folkestone	Coope (1980)
27. Great Yarmouth	Jones (1976)
28. Hampstead Heath	Girling and Greig (1977, 1985)
29. Hasholme	Holdridge (1987)
30. Hen Domen	Greig et al. (1982)
31. Hereford	Girling (1985a), Kenward (1985a)
32. Hull	Kenward (1977, 1979b)
33. Knap of Howe, Orkney	Kenward (1983)
34. Leicester	Girling (1981)
35. Lindow	Girling (1986), Skidmore (1986)
36. Malverns (Midsummer Hill)	Osborne (1981b)

Table 7.1 (*Continued*)

Site	Reference(s)
37. Messingham	Buckland (1984)
38. Northampton	Keepax et al. (1979), Robinson (1983)
39. North Ferriby	Buckland et al. (1990)
40. Oxford	Robinson (1979b, 1981b, 1984, 1986)
41. Papcastle	Kenward et al. (1988)
42. Porth Meare Cove	Osborne (1976)
43. Sandtoft	Samuels and Buckland (1978))
44. Sewerby	Girling (1985b)
45. Somerset Levels	Coles (1987), Girling (1976a, 1976b, 1977, 1978, 1980, 1982, 1984)
46. Southampton	Buckland et al. (1976), Kenward and Allison (1987), Kenward and Girling (1986)
47. Southwark	Girling (1979)
48. St. Ola, Orkney	Smith (1981)
49. Tattershall Thorpe	Chowne et al. (1986)
50. Taunton	Osborne (1984)
51. Thorne Moor	Buckland (1979), Buckland and Kenward (1973)
52. Towcester	Girling (1983)
53. Westward Ho	Girling and Robinson (1987)
54. Whitton	Osborne (1981c)
55. Wilcote	Robinson (1978)
56. Wilsford	Osborne (1969, 1986, 1989)
57. Worcester	Colledge and Osborne (1980), Osborne (1981d)
58. York	Buckland (1973, 1974, 1976a), Buckland et al. (1974), Buckland and Kenward (1973), Hall and Kenward (1976), Kenward and Williams (1979)

by the nobility (including that much-maligned conservationist, William the Conqueror), but even these forests were constantly under pressure (Buckland, 1979).

Bronze and Iron Age trackways have been excavated in the marshy region known as the Somerset Levels (Fig. 7.2, No. 45). Insect faunas from these sites were described by the late Maureen Girling (Girling, 1976a,b, 1977, 1978, 1980, 1982, 1984; Coles, 1987). Prehistoric farmers tried to develop agriculture in boggy lowlands with the aid of an extensive series of wooden trackways, but these measures eventually failed and the farms were abandoned.

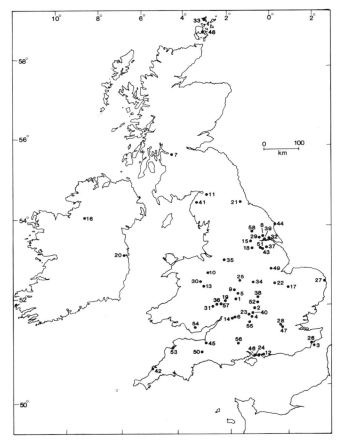

Figure 7.2. Map of the British Isles showing locations of archaeological sites from which insect fossils have been analyzed. Site numbers are keyed to the list of sites in Table 7.1.

A Bronze Age well at Wilsford, Wiltshire (Fig. 7.2, No. 56), was investigated by Osborne (1969, 1986, 1989). The samples of organic detritus came from the bottom of the well (a depth of 30 m); 23 liters of sample produced a beetle fauna of 2600 individuals in 138 taxa and 27 families! The insect horizon dates to 3330 yr B.P. and suggests a climate similar to the modern one, or slightly warmer (Osborne, 1969). Contrary to archaeological expectations that this would prove to be a ritual pit, the insects indicate that it was an everyday farmer's well used for watering cattle (based on a superabundance of dung beetles in the fossil fauna). Quantities of cut grasses were brought near the well, probably as fodder or bedding, as suggested by the presence of beetle species that live in decaying grasses. The presence of wood-boring beetles suggests that timbers were erected over the mouth of the well.

Coope and Osborne (1968) described an insect fauna from a Roman well at Barnsley Park in central England. This fauna included stored product pests.

Iron Age records abound in Britain, and the occupation of the Roman Army also left its mark on the landscape. The Romans built many forts and other strongholds, as well as houses, baths, and other buildings. In one of the northern outposts at Bearsden, Scotland (Fig. 7.2, No. 7), insect fossils from a ditch included stored product pests associated with grain (Dickson et al., 1979). These pests were also abundant in samples taken from a Roman warehouse at York (Fig. 7.2, No. 58) (Hall and Kenward, 1976; Kenward and Williams, 1979). At Bearsden, they were found in cereal debris, which may have entered the ditch as sewage, presumably after both the grain and the pests had been consumed by Roman soldiers.

Another curiosity from the Iron Age is the Lindow Man, a corpse preserved in a *Sphagnum* bog in Cheshire (Fig. 7.2, No. 35). The insects associated with this "bog man" are a part of the forensic evidence of how he died (Girling, 1986; Skidmore, 1986).

Reconstructions of medieval life at Anglo-Danish sites in York have included detailed research on insect fossils from sediments taken beneath Lloyd's Bank and subsequently at Coppergate (Buckland, 1973, 1974, 1976b; Buckland et al., 1974; Addyman et al., 1976; Hall and Kenward, 1976, 1990; Kenward et al., 1978, 1986; Hall et al., 1983; O'Connor et al., 1984; Kenward, 1988; Addyman, 1989). Denford (1978) has also examined fossil mites from the Coppergate site. More than 5 m of waterlogged sediments have accumulated in York from Roman through medieval times, leaving layer upon layer of refuse. All of the insect assemblages from York reflect human habitations. The preservation is excellent because of the waterlogging. Based on the quantity and quality of debris, initial research suggested that the Romans who occupied York were somewhat more hygienic than the subsequent Anglo-Danish inhabitants. Further work has cast the Anglo-Danish in a better light (Hall and Kenward, 1990). Nevertheless, the overwhelming impression from the insect evidence is one of squalor in Anglo-Danish time. Rush and reed flooring materials were rarely changed; a well-developed compost fauna lived in every building studied. This life-style was not unique to York, however. Coope (1981) documented the same type of floor debris in a Viking house from Dublin (Fig. 7.2, No. 20). He hypothesized that the heat from fermenting plants on floors might have helped warm the house in winter. In contrast to these findings, a group of Saxon pits at Southampton (Fig. 7.2, No. 46) indicate that mammal dung was removed from buildings and buried outside (Buckland et al., 1976; Kenward and Girling, 1986; Kenward and Allison, 1987).

At a tannery site in York dating to A.D. 1000 the remains of thousands of flies (houseflies and biting stable flies) were found. The investigators also found chicken feathers and eggshells (Buckland et al., 1974). Chicken manure was probably used in tanning hides. Among the manure a beetle was found that is common on the dung-covered floors of chicken coops. There was no evidence of stored grains at the tannery, just mealworms that were probably brought in with some flour.

In a York house, a fossil insect assemblage provides evidence of heath plants used as bedding, a practice still seen in parts of Ireland. Unfortunately for the would-be slumberers, the heath insect fauna includes numerous biting ants (Buckland et al., 1974).

Timbers in York buildings were infested with deathwatch and powderpost beetles. Although little is left of the original walls, some clues to their construction are found in the beetle remains. The presence of dung beetles in combination with willow stem–eating beetles indicates the use of wattle and daub architecture: dung was used to make daub and the willow-eating beetles may be traced to the wattle. Nettle-feeding beetles were also found in the house. Buckland et al. (1974, p. 31) suggested that "perhaps they wandered in from the back yard" of this bijou residence.

The evidence from beetle species brought in from natural habitats (e.g., riverbanks, peat moss, heather) suggests that the climate of medieval York was similar to that in modern times (O'Connor et al., 1984). The apparent affluence of the people based on their artifacts (jewelry, imported pottery) contrasts sharply with the squalor inferred from the biological evidence. Thus the entomological studies have modified our image of medieval York considerably from that based on earlier archaeological studies.

H. K. Kenward (written communication, 1990) has also been working on an early Christian archaeological site at Deer Park Farms, Antrim, Northern Ireland. This work is forcing a reevaluation of the ways in which the characteristic urban fauna developed, because species have been found at this isolated, rural site that were formerly regarded as slow colonizers, needing long-lived settlements in order to become established. Clearly, much remains to be learned about how insects have become adapted to anthropogenic environments. The work of these British paleoentomologists constitutes a first major step.

FORENSIC PALEOENTOMOLOGY

Taxonomic and ecological data gathered by forensic entomologists for use in criminology have also been put to use in archaeological contexts. Forensic entomological research has documented the insect faunal succession on human

Figure 7.3. Map of Western Europe showing locations of archaeological sites from which insect fossils have been analyzed. Site numbers are keyed to the list of sites in Table 7.2.

and other corpses, allowing more precise reconstructions of the timing of deaths, modes of burial, and other circumstances surrounding death and burial (Erzinclioglu, 1983; Smith, 1986). Successive "waves" of insect groups are attracted to corpses during the various stages of decomposition. These carrion insects include a variety of maggot (fly larvae) species and both larvae and adults of beetles in more than 16 families, including the Carabidae, Silphidae, Leiodidae, Staphylinidae, Histeridae, and Dermestidae. The same insects have been attracted to carrion for many thousands of years, so the fossil insects found with ancient human and domesticated animal remains have likewise provided good evidence on the timing and mode of ancient deaths and burials.

STUDIES IN WESTERN EUROPE

In the Netherlands, Hakbijl (1989) has investigated insect remains from an early Iron Age site at the Assendelver Polders, west of Amsterdam (Fig. 7.3, No. 2). More than 175 beetle and other insect taxa were identified, including

Table 7.2

European archaeological sites with insect fossil analyses

Site	Reference(s)
1. Ageröd V, Sweden	Lemdahl (1982)
2. Assendelver Polders, The Netherlands	Hakbijl (1989)
3. Champreveyres, Switzerland	Coope and Elias (unpublished data)
4. Falun, Sweden	Hellqvist and Lemdahl (1990)
5. Fiave, Italy	Osborne (1985)
6. Halmstad, Sweden	Lemdahl and Thelaus (1989)
7. Holt, Iceland	Sveinbjarnardóttir (1983), Buckland et al. (1991)
8. Ketilsstadir Farm, Iceland	Buckland et al. (1986a)
9. Kuppajarvi, Finland	Salonen et al. (1981)
10. Leeuwarden, The Netherlands	Schelvis (1987)
11. Lower Rhine River, The Netherlands	Klink (1989)
12. Lund-Apotekaren, Sweden	Lemdahl (1991c)
13. Oldeboorn, The Netherlands	Schelvis (1990a)
14. *Oskarshamn* shipwreck, Sweden	Lemdahl (1991e)
15. Oslo, Norway	Kenward (1980)
16. Rhineland, Germany	Koch (1970, 1971)
17. Santorini, Greece	Panagiotakopulu and Buckland (1991)
18. Skateholm, Sweden	Lemdahl (1988a)
19. Storaborg, Iceland	Perry et al. (1985)
20. Svendborg, Denmark	Jorgensen (1986)
21. Uppsala, Bryggaren, Sweden	Lemdahl (1991d)
22. Wilhelmshaven, Germany	Lemdahl (1990)

substantial numbers of salt marsh and other saline environmental indicator species. Excavations of sandy floor samples yielded the largest percentage of halobionts. Most of these insects show a preference for some type of water-land transitional environment, such as the strand of a salt marsh creek or the edge of a tidal pond. Their presence suggests that house floors in the settlement included layers of sand from salt marsh levees. The insect remains also led Hakbijl to conclusions about the gathering and indoor use of various food items, and, although other evidence indicates that the houses were kept relatively free of rubbish, poor sanitary conditions are suggested by the presence of human fleas, abundant houseflies, and stable flies.

Schelvis (1987, 1990a,b) uses fossil oribatid mites to interpret the environment of archaeological sites. His work on the Dutch medieval site of Oldeboorn (Fig. 7.3, No. 13) documented types of local landscapes, ranging from dry litter- and moss-covered soil through perennially wet moorland. He has also investigated medieval mite fossils from Leeuwarden (Fig. 7.3, No. 10) (Schelvis, 1987).

Klink (1989) compiled a large body of insect fossil data concerning anthropogenic changes in the lower Rhine River in the Netherlands (Fig. 7.3, No. 11). A total of about 15,000 insect remains in 167 taxa were recovered from 52 sediment core samples from the Rhine. Most were aquatic larvae, including flies, caddisflies, and stoneflies. Klink's study indicates that the insect fauna of the Rhine has undergone a complete change during historic times, with 80–100 insect species disappearing during the last few centuries. Numerous deformed head capsules of aquatic fly larvae were found. Such deformities are thought to have been caused by the dumping of heavy metal compounds into the river during the last 200 years.

Hakbijl (1987) has also delved into insect fossils uncovered as part of marine archaeological studies. The wrecked hull of the Dutch East Indies ship *Amsterdam,* found in the English Channel near Folkestone (Fig. 7.2, No. 3), yielded a container of "Spanish fly," the ground-up exoskeletons of the blister beetle (Meloidae), *Lytta vesicatoria.* However, closer examination showed that the shipment had been adulterated with similarly colored bodies of the green chafer (Scarabaeidae), *Cetonia aurata.* Apparently the adulteration of "controlled substances" with cheap fillers has been common practice for quite some time.

The Swiss lake village of Champreveyres at Neuchâtel (Fig. 7.3, No. 3) has been the subject of an interdisciplinary archaeological/environmental study, beginning with the Magdalenian occupation during the late glacial interval. The paleoenvironmental work was based on lake sediments adjacent to the occupation site (Coope and Elias, unpublished data; Egloff, 1987). Insect fossils were described from sediments spanning the interval 12,670–11,820 yr B.P. The site yielded 171 beetle taxa in 36 families. The only evidence of human disturbance is from a mixture of temperate and arctic beetles in one horizon. Perhaps people mixed the sediments through digging, although such disturbances were not evident in the stratigraphy of the site. No synanthropic insects were found, and the Paleolithic inhabitants seem to have had little impact on local landscapes.

Osborne (1985) studied a small insect fauna from a Bronze Age lake dwelling at Fiave, Italy (Fig. 7.3, No. 5). He found the remains of aquatic insects that lived in still waters at the site, as well as stored product pests and dung beetles that may have fed on the dung of domesticated animals.

Elsewhere in Europe, insects have been studied from medieval sites at Oslo, Norway (Fig. 7.3, No. 15) (Kenward, 1980); Svendborg, Denmark (Fig. 7.3, No. 20) (Jorgensen, 1986); and the Orkney Islands (Fig. 7.2, Nos. 33 and 48) (Kenward, 19883; Smith, 1981). More extensive work has been done in Germany, where Koch (1970, 1971) has studied insects from Roman and medieval sites in the Rhineland (Fig. 7.3, No. 16) and Lemdahl (1990) has investigated insects from an Iron Age settlement near Wilhelmshaven (Fig. 7.3, No. 22).

In Greece, Panagiotakopulu and Buckland (1991) have studied fossil insects from a late Bronze Age site at Santorini, on the island of Thera (Fig. 7.3, No. 17). The fossil evidence indicates the establishment of field and stored grain pests in the Aegean region by about 3500 yr B.P. The cereal pest *Rhyzopertha dominica* (Coleoptera: Bostrichidae) probably originated in Africa and arrived at Santorini through trade with Crete and Egypt.

In Sweden, Mesolithic and early Neolithic sites at Skateholm in the south (Fig. 7.3, No. 18) have yielded insect faunas indicative of natural lacustrine, riparian, and coastal marine environments (Lemdahl, 1982, 1988a; Lemdahl and Thelaus, 1989). The insect fauna is consistent with the development of a lagoon at the site, caused by a marine transgression. Roughly contemporaneous insect assemblages from the Ageröd V bog site (Fig. 7.3, No. 1) are likewise indicative of natural environments adjacent to a village. The fauna includes thermophilous species that have subsequently been extirpated from Sweden. Insect assemblages from the fifteenth-century Halmstad site in southwestern Sweden (Fig. 7.3, No. 6) also include species that have since been regionally extirpated. These shifts in distribution have been attributed to either changes in land use patterns or changing climate.

Lemdahl (1991c,d) also studied insect assemblages from medieval sites in the cities of Lund and Uppsala (Fig. 7.3, Nos. 12 and 21). Small faunas were recovered from refuse layers, wells, and barrels. The most recent fossils studied from Swedish archaeological sites come from Falun (Fig. 7.3, No. 4), where 300-year-old buildings have been excavated. Soil from within the houses yielded 43 insect taxa (Hellqvist and Lemdahl, 1990). The shipwreck of the frame-timbered vessel *Oskarshamn* (Fig. 7.3, No. 14) also yielded medieval insect fossils (Lemdahl, 1991e), although that research is still in progress.

STUDIES IN THE NORTH ATLANTIC REGION

In southern Iceland, Buckland et al. (1986a) provided records of pre- and postsettlement insect faunas from trenches dug through peat associated with an archaeological site at Ketilsstädir Farm (Fig. 7.3, No. 8). The Landnám tephra was deposited at the time of the first Viking landings, about A.D. 900 and hence provides a convenient marker horizon. Landnám is the Old Norse term for the taking or settling of land. Altogether, 46 beetle samples indicate a fauna of a few species that became more diverse as humans modified the landscape (especially through the draining of bogs and the overgrazing of hillsides by sheep). Among the synanthropic species found were a dung beetle, *Aphodius lapponum,* that feeds on the dung of large herbivores, and several species associated with stored hay.

Another study in southern Iceland focused on insect faunas from a farm at Holt (Fig. 7.3, No. 7) (Buckland et al., 1991; Sveinbjarnardóttir, 1983). This paper discussed the ecological effects of the Norse colonization of Iceland, which began in the ninth century A.D. These impacts were investigated through comparisons between the pre- and post-Landnám insect faunas and floras of this and other sites. The farm at Holt was settled early and appears to have been continuously occupied for more than 1000 years. By sampling organic deposits from a series of ditches, the authors were able to obtain fossil assemblages in stratigraphic sequence. The pre-Landnám insect fauna from Holt was composed of subarctic species. The post-Landnám fauna included several synanthropic insects, including *Omalium excavatum, Xylodromus concinnus, Xylodromus depressus, Atomaria* cf. *apicalis, Mycetaea hirta, Corticaria elongata,* and *Typhaea stercorea.* The data obtained from the Holt study provided the basis for a regional synthesis of paleoenvironments. This showed that the pre-Landnám, subarctic biota of Iceland represented a fragile ecosystem that fared very poorly after the Norse settlement of the island. Human occupation has been ecologically devastating to much of Iceland, and this devastation began with the earliest settlers, as indicated by the fossil study.

Perry et al. (1985) discovered a mixture of ecological groups in the fossil insect fauna from the late medieval farm at Storaborg, Iceland (Fig. 7.3, No. 19). The authors used cluster analysis to help refine their interpretation of the faunal elements. Two principal groups were distinguished: a midden component and a house floor component. Taken together, the evidence of insects and plant macrofossils pointed to domestic areas that had a warm, foul, well-trod covering of cut vegetation containing large quantities of decomposing garbage. In contrast, the outdoor midden-type layer, exposed to the elements, had a cool, less foul, less stable environment.

Following the colonization of Iceland, the Vikings proceeded west to establish settlements on Greenland. Cereal agriculture was precluded by the severe climate, and by about A.D. 1350 the settlements in western Greenland were abandoned (Fredskild, 1988). The work of archaeologists and paleoecologists has given us a glimpse of what life was like for these hardy folk. Archaeological investigations of

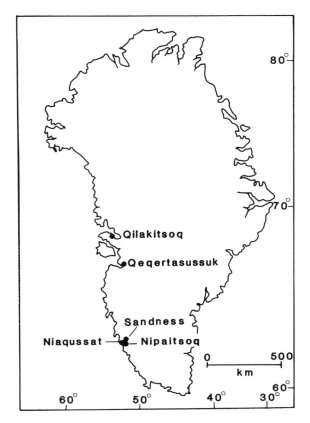

Figure 7.4. Map of Greenland showing locations of archaeological sites from which insect fossils have been analyzed.

the remains of farms from Niaqussat and Nipaitsoq (Fig. 7.4) on the southwestern coast were part of the Inuit-Norse Project (McGovern et al., 1983). The farms date from circa A.D. 1000–1350. Insect remains in detritus layers on floors of farm buildings indicate refuse dumping from the living rooms and barns. Several species associated with damp, moldering hay died out in Greenland when the Viking colonies were abandoned. The middens also contained some native Greenland beetles, including subarctic ground beetles and weevils. In living rooms, abundant byrrhids provide evidence that mosses were brought in, perhaps to fill bedding. Abundant remains of fly puparia (species often associated with human feces) were also found in living rooms. Such flies are found in many unhygienic environments as well as in latrines and sewers. McGovern et al. (1983, p. 706) concluded that "in a place where it was essential to conserve as much heat as possible, it is probable that visits to the outside world were kept at an absolute minimum during winter, and the use of one room for defecation, even if it was not strictly a latrine, is not surprising." The living rooms also showed evidence of rotting meat and feces, covered over by fetid hay.

Historians have pieced together accounts of life in Norse Greenland, although few firsthand accounts exist and even fewer discuss the squalid living conditions. Life in eighteenth-century Iceland may not have been too different, as the following account relates: "The floor was bare, made of packed earth which turned to mud when the roof leaked during rainstorms. The walls were covered with a grayish form of mould, and a greenish, slimy liquid dripped constantly down them, particularly in winter" (Jónsson, 1892–1894, p. 107). It is tempting to equate the squalor with a low standard of living, but the decaying vegetation on the floor probably insulated against the permafrost. The relationship between poor hygiene and poor health had not been established in medieval times; Greenland farms were no more squalid than city dwellings in York.

Böcher's (1988) analysis of fossils from the Inuit archaeological site of Qeqertasussuk (Fig. 7.4), occupied from about 4400–3200 yr B.P., yielded insect assemblages reflecting faunas more diverse than those occupying the site today. Several species of beetles in the fossil assemblages have modern ranges significantly farther south in Greenland. These include the ground beetle *Trichocellus cognatus* and the rove beetles *Micralymma brevilingue* and *Atheta islandica*. Based on the insect assemblage data, the climate at 3500 yr B.P. at Qeqertasussuk was warmer than at present.

The only arthropod indicators of human disturbance at the site were abundant remains of litter-inhabiting sheet-web spiders (Linyphiidae) and abundant fly pupae, suggesting that maggots fed on decaying meat or other garbage.

STUDIES FROM OTHER OLD WORLD REGIONS

Other archaeoentomological studies have been performed at sites scattered throughout the Old World. A number of studies have been made on insects associated with Egyptian tombs (e.g., Solomon, 1979) and with tombs from ancient Chinese dynasties (Chu and Wang, 1975). In Malawi, fossil termite mounds have been investigated because of their association with stone artifacts (Crossley, 1984).

STUDIES OF INSECT PARASITES FROM ARCHAEOLOGICAL SITES

The mummified corpses of six Inuit women and two children, dating to the fifteenth century A.D., have been described from the Qilakitsoq site in western Greenland (Fig. 7.4) (Hansen, 1989). Many lice and their eggs were found on these mummies, especially on one woman. Some lice were even in her intestines, indicating their ingestion (Bresciani et al., 1983, 1989). Like the human corpses, the lice were essentially freeze-dried and in an excellent state of preservation. However, the discovery of lice remains associated with humans or their habitations is not isolated to Greenland. Lice have also been found in coprolites of Paleoindian age in Utah (Fry, 1976) and in sixteenth-century Aleutian Island mummies (Horne, 1979). Evidence from the abandoned farms in Greenland indicates that the Vikings also had lice, but they probably brought them from Scandinavia, since the Vikings had little contact with the Inuit.

Norse farms on Greenland, Iceland, and Orkney have yielded flea remains. The Norse Greenland farms revealed remains of *Pediculus humanus humanus* (Sveinbjarnardóttir and Buckland, 1983; Sadler, 1990), the louse that is found on the trunk of humans. It requires temperatures of 29–30°C, whereas the head louse, *Pediculus humanis capitis,* can withstand lower temperatures. These two parasites, in addition to the crab louse, *Phthirus pubis,* are known in antiquity. Lice infestations brought on by squalor and lack of personal hygiene have been ubiquitous until very recently. They may contribute greatly to the spread of disease, but this association was not well known in the Middle Ages. Poor living conditions were seldom commented on in medieval literature (why state the obvious?), except that authors from one country would sometimes comment on unsanitary conditions in another. The economic failure of the Norse farms on Greenland was probably the cause of their demise, but louse-borne disease cannot be ruled out.

Fleas have also been found in Old World archaeological sites, beginning with Neolithic remains from the island of Orkney (J. P. Sadler, personal communication, 1992), through Roman-age archaeological sites at York (Hall and Kenward, 1990), and on through medieval assemblages from York, Dublin, and Norse Greenland.

Synanthropic arthropods have also been found in archaeological sites in South America. Baker (1990) analyzed fossil mites from the guts of human mummies in Tarapaca, northern Chile. *Lardoglyphus robustisetosus* was identified from a desiccated mummy dated between 2400 and 2000 yr B.P. Baker suggested that mites of this species were ingested with food. The gut contents in which the mites were found consisted of a high-protein pemmican, which, as a stored product, would have attracted *Lardoglyphus* mites.

INSECT FOSSILS IN NEW WORLD ARCHAEOLOGICAL SITES

Most insect fossil studies from New World archaeological sites deal with contemporaneous faunas from more or less natural environments in the vicinity

Figure 7.5. Map of North America, showing locations of archaeological sites from which insect fossils have been analyzed. Site numbers are keyed to the list of sites in Table 7.3.

of the archaeological site. In this respect, New World studies differ from most Old World studies, which concern insect assemblages from ancient human habitations. A few American studies have archaeological connections, however. Graham (1965) examined insect fossils from a number of localities in the famous Anasazi Indian sites at Mesa Verde, Colorado (Fig. 7.5, No. 16). A sequential study of insect assemblages, spanning the interval from the Basketmaker culture through the Pueblo culture, indicates that the synanthropic insect fauna remained virtually unchanged. Stored product pests of corn were found both in dried corn samples and in human coprolites.

Although human parasites such as lice were not found in occupation sites at Mesa Verde, louse eggs were observed on mummified corpses. Some buried human remains were heavily infested with various kinds of blowflies (Diptera: Calliphoridae), suggesting that the fly eggs were laid while the body was being prepared for burial, presumably during the summer months (the flies' active period). Other corpses and skeletons were devoid of fly larvae; these were probably buried in winter.

Table 7.3

North American archaeological sites with insect fossil analyses

Site	Reference(s)
1. Aubrey Spring, Texas	Elias, unpublished data
2. Bamert Cave, California	Nissen (1973)
3. Big Bone Cave, Tennessee	Faulkner (1991)
4. Bighorn Basin, Wyoming	Chomko and Gilbert (1991)
5. Bonnet Plume Basin, Yukon Territory	Hughes et al. (1981)
6. Danger Cave, Utah	Fry (1976)
7. Dirty Shame Rockshelter, Oregon	Hall (1977)
8. False Cougar Cave, Montana	Elias (1990b)
9. Glen Canyon, Utah	Fry (1976)
10. Hogup Cave, Utah	Fry (1976)
11. Huntington Canyon, Utah	Elias (1990b)
12. Lamb Spring, Colorado	Elias (1986), Elias and Nelson (1989), Elias and Toolin (1989)
13. Leavenworth, South Dakota	Gilbert and Bass (1967)
14. Lovelock Cave, Nevada	Heizer (1970)
15. Lubbock Lake, Texas	Elias and Johnson (1988)
16. Mesa Verde, Colorado	Evans and Baldwin (1977), Graham (1965)
17. Ocampo Caves, Mexico	Callen (1970)
18. Old Crow, Yukon Territory	Matthews (1975a), Matthews et al. (1990a), Morlan and Matthews (1978, 1983)
19. Plainview, Texas	Holliday et al. (1993)
20. San Fernando de Vellicata, Baja California	Essig (1927)
21. San Vincente Ferrer, Baja California	Essig (1927)
22. Snowflake, Arizona	Hevley and Johnson (1974)
23. Tehuacan Valley, Mexico	Callen (1967)

Bark beetle evidence from timbers cut for use in cliff dwellings reveals a number of aspects of this activity. The Anasazi cut down trees from the top of the mesa and stripped them of bark before placing them in walls. This bark removal probably took place in July or early August. At this time phloem-boring beetles enter the wood and thereafter feed only on wood. After a year or more, these beetles emerge as adults, cutting a characteristic exit hole. These exit holes were for the most part missing from the Mesa Verde timbers, indicating that the bark had been removed in midsummer while the larvae of phloem borers were still mining between the bark and the wood. Furthermore, it appears that the Indians mostly felled the wood in the early summer and then left it on the ground, allowing the phloem-feeding bark beetles to perform the arduous task of loosening the bark.

Dermestid larval exuviae (the skins cast from the larvae as they pupate) were identified by Evans and Baldwin (1977) from a cake of salt recovered from an Anasazi jar at Oak Tree House at Mesa Verde. Apparently the insect remains had been mixed accidentally with loose, granular salt as it was moistened and molded into a cake.

The study of ancient human feces, preserved in dry caves as coprolites, has yielded some interesting information on insects and their interactions with Amerindians. Some arthropods were consumed as food. Others, such as fleas, lice and mites, were external parasites that were either accidentally or deliberately consumed.

Stiger (1977) found the remains of cicadas and grasshoppers in Anasazi coprolites from Mesa Verde. He suggested that insects became increasingly important in the Anasazi diet through time and that regional environmental change may have played a role in this sequence. As shrub-grassland became dominant, the numbers of grasshoppers would have increased. Evidence of increased numbers of turkey bones in coprolites also led Stiger to suggest that domesticated turkeys were used to control grasshoppers on cultivated lands.

A number of mites found in human coprolites from Mesa Verde (Morris, 1986) have not been identified to the species level, but some belong to the genus *Trombicula,* known as chiggers in the American southwest. The larvae of these mites parasitize humans by burrowing beneath the skin to feed on blood and interstitial fluid.

Paleoindian coprolites from the Dirty Shame Rockshelter, Oregon (Fig. 7.6, No. 7), have been dated at 9500 yr B.P. Some contained abundant remains of red ants (Hymenoptera: Formicidae: *Formica* sp.); others were comprised almost completely of indigestible parts of the termite species *Reticulitermes* cf. *tibialis* (Isoptera) (Hall, 1977). Termites comprised 78% of one coprolite, which indicates that Paleoindians sometimes made a complete meal of termites. Consumption of insects by Late Prehistoric Amerindians from western North America was documented by the European historians who first made contact with Amerindians and corroborated by fossil evidence (Stewart, 1938; Madsen and Kirkman, 1988). For example, Nissen (1973) found remains of craneflies (Diptera: Tipulidae) in Amerindian coprolites from Bamert Cave, California (Fig. 7.6, No. 2).

In the Great Basin region of western North America, many dry caves contain human coprolites. In Utah, Fry (1976) analyzed coprolites found in Danger and

Hogup Caves, as well as those from caves in the Glen Canyon region (Fig. 7.6, Nos. 6, 10, and 9). A 2000-year-old coprolite from Danger Cave contained an egg case (nit) of *Pediculus humanus,* the human louse, still attached to a human hair. A 5000-yr-old coprolite from Danger Cave contained the complete exoskeleton of a mosquito. The ingestion of the mosquito must surely have been accidental; it would take a great many mosquitos to make a meal! Fry (1976) also found that human mummies from caves in Glen Canyon had louse-infested scalps.

Heizer (1970) summarized the results of insect fossil studies from coprolites in Lovelock Cave, Nevada (Fig. 7.6, No. 14). Some specimens, dating between 1200 and 150 yr B.P., contained the exoskeletons of the large predaceous diving beetle *Cybister.* The heads were missing, suggesting that, being less palatable, they were removed before the beetles were swallowed.

Human coprolites dated at about 2000 yr B.P. have been studied from Big Bone Cave, Tennessee (Fig. 7.6, No. 3). Faulkner (1991) examined eight samples in order to reconstruct diet, disease, and life-style. In addition to food items in the feces, fleas, weevil larvae, and fly fragments were found. The weevil larvae were probably derived from weevil-infested grain in the diet. The weevil species is in a group known to attack seeds of legumes and other plants that could be stored for food. The fleas found were woodrat (*Neotoma*) fleas and may represent postdepositional invasion of the feces in the cave. On the other hand, they may have been eaten by the Indians. Noninsect parasites, such as pinworm eggs, roundworm eggs, hookworm eggs, and *Giardia* cysts, were also found in the fecal samples.

In Mexico, coprolites from cave deposits spanning much of the Holocene have been investigated for insect remains. Callen (1970) studied coprolites from caves near Ocampo, Tamaulipas (Fig. 7.6, No. 17). Some of the coprolites were riddled with exit holes and pupal cases from blow flies and flesh flies. Larval exuviae of the latrine fly, *Fannia scalaris,* were also abundant, in addition to the remains of the housefly, *Musca domestica,* and fungus gnats. Several species of dermestid beetles were identified. Among these, the larvae of the odd beetle, *Thylodrias contractus* (Dermestidae), is the most remarkable. This beetle, a scavenger on dried animal material and fur, is thought to have been carried on dried animal skins by humans migrating over the Bering Land Bridge into the New World (Callen, 1970).

Another set of sites with preserved coprolites was discovered in the Tehuacan Valley of central Mexico (Fig. 7.6, No. 23) (Callen, 1967). Insect fossils have been identified from coprolites ranging in age from 8700 to 450 yr B.P. Early coprolites (>5400 yr B.P.) contained ant heads, dung beetle larvae, caterpillars, and ticks. Coprolites dating from 2900 to 2200 yr B.P. contained dung beetles, fruitfly (*Drosophila*) larvae, latrine fly larvae, and chewing lice (Mallophaga). The lice were probably bird parasites, accidentally ingested by people. Coprolites dating from 2200 to 1300 yr B.P. contained sucking lice (*Anoplura*), fleas, ticks, flies, beetles, ants, and pseudoscorpions. The beetles were sufficiently broken up to demonstrate that they had been chewed. Beetle exoskeletons formed the bulk of several individual coprolites. Some of the fossil ants resembled honey ants (*Myrmecocystus* sp.), whose abdomens store honeydew, a sugary excretion harvested by the ants from the abdomens of aphids. These ants may have been deliberately consumed for their sweet taste.

In Arizona, insect remains have been extracted from debris middens in rooms in a prehistoric period pueblo near the modern town of Snowflake (Fig. 7.6, No. 22) (Hevley and Johnson, 1974). A small assemblage of darkling beetles, ants, and ichneumon wasps was found. The authors speculated that these scavenging insects were attracted to food resources in the pueblo and that the parasitic wasps then preyed on the scavengers. They noted that insects are found in many such midden deposits but that this line of research had yet to be pursued.

Indian pueblo buildings have traditionally been made of adobe, a mixture of clayey mud and straw, sun-baked into bricks. The bricks are made into walls that are then plastered over with more mud to form a smooth finish. When the Spanish came to the American southwest, they adopted the use of adobe in making their forts, houses, and mission churches. Insect fossils retrieved from old adobe walls from Spanish mission churches at San Fernando de Vellicata and San Vincente Ferrer (Fig. 7.6, Nos. 20 and 21) in Baja California (Essig, 1927) comprise species that lived in the stream bank mud that was used to make the adobe, mud wasps that made nests in the fresh adobe mud, and stored product pests, including the granary weevil (*Sitophilus granarius*) and the rice weevil (*Sitophilus oryzae*). Essig speculated that these pests were introduced into the mission communities by priests who brought them in bags of contaminated grain.

The remains of fly maggots and their pupal exuviae have been used in some New World sites to help determine mode and timing of burial. Gilbert and Bass (1967) studied fly remains from a Late Prehistoric Arikara Indian burial site at Leavenworth, South Dakota (Fig. 7.6, No. 13). Exuviae of many flies were found, placing the time of burial between late March and mid-October (the flies' active season in South Dakota).

In the Bighorn Basin of Wyoming (Fig. 7.6, No. 4), a Late Prehistoric bone pit was analyzed by Chomko and Gilbert (1991). The contents of a bone-filled pit were thought to be the remains of a meal cooked in a skin container; after cooking, the residue would have been dumped into the pit and covered with a slab of stone. However, abundant remains of blowfly puparia found at the bottom of the bone mass provide evidence that the bone mass was formed outside the pit, because the flies could not have burrowed that deeply through bones to lay their eggs at the bottom. Flies are attracted to rotting meat, so the contents of the container must have been exposed a few days before burial. Development of maggots and maturation through several instars until pupation requires at least 21 days in warm weather. So the contents were exposed at least three weeks before being dumped into the pit. This series of events probably took place between June and September, the active season for blowflies in Wyoming. If it had occurred in winter or spring, secondary carrion beetles such as dermestids would have been attracted to the bone mass after it thawed in spring. These beetles feed on dried carrion and are thus predisposed to become pests in dried meats and animal skins.

INSECTS FROM NONANTHROPOGENIC DEPOSITS AT ARCHAEOLOGICAL SITES

Insect fossils from nonanthropogenic environments (i.e., undisturbed, natural habitats) provide data useful in building an environmental framework within which to place archaeological interpretations of a site. Insects remains, along with other biological proxy data, enable inferences on climate, soil, and regional vegetation to be drawn.

I have studied a number of insect fossil assemblages from this type of deposit, including samples from the Aubrey Spring Clovis site near Denton, Texas (Fig. 7.6, No. 1). Abundant insect assemblages were recovered from pond sediments. The insects range in age from 14,000 to 13,300 yr B.P. This interval predates a Clovis Paleoindian occupation on a hill adjacent to the ancient pond. Unfortunately, the pond sediments dating to the time of Clovis occupation contained few insects.

The older sediments yielded an abundant, diverse fauna indicative of open ground environments surrounding a sedge-lined pond. The modern distributions of the insects suggest a climate with summers about 10°C cooler than the present ones at 14,200 yr B.P., warming 2°C by 13,500 yr B.P. This scenario agrees with a contemporaneous grassland pollen spectrum but adds some additional climatic definition.

I have also examined insect fossils from a Clovis archaeological site at False Cougar Cave, Montana (Fig. 7.6, No. 8) (Elias, 1990b). The insects were recovered from cave floor sediments that also contained Paleoindian artifacts. The age of the fossil insect sample is 10,500 yr B.P. The fauna is small but provides an environmental reconstruction that is internally consistent. The insects in the assemblage came from outside the cave (there are no cave dwellers). Based on the insects, the environment was very similar to modern conditions by 10,500 yr B.P.

At the Lubbock Lake site in west Texas (Fig. 7.6, No. 15), I have sampled insects from an organic horizon including bison bones butchered by Clovis Indians. The age of this sample is circa 10,300 yr B.P. Only a few beetle species were identified (Elias and Johnson, 1988). One of these is the ground beetle species *Calosoma porosifrons,* a caterpillar hunter. Its modern range is limited to mountain slopes in the state of Durango, Mexico. The climate there is cooler and wetter than that of modern-day Lubbock. Younger (nonwaterlogged) strata at Lubbock Lake apparently are oxidized too heavily for good preservation of insects.

At the Plainview site in west Texas (Fig. 7.6, No. 19), I recovered insect fossils from organic horizons in a sand and gravel pit. The samples range in age from 10,000 to 9000 yr B.P., contemporaneous with Paleoindian archaeology at Plainview but not in direct association with artifacts. A limited beetle fauna shows climatic conditions similar to modern ones.

Lamb Spring, Colorado (Fig. 7.6, No. 12), is a multicomponent archaeological site, with a basal bone bed including many extinct megafaunal mammals. Stanford et al. (1982) described possible Pre-Clovis artifacts at the site. A radiocarbon date on mammoth bone was 13,100 ± 1000 yr B.P.; plant macrofossils yielded an age of 12,750 ± 150 yr B.P. I sampled insect fossils from the bone bed clay horizon (Elias, 1986). This fauna was ecologically mixed, including both prairie and alpine tundra species. Accelerator mass spectrometry (AMS) dating of fossil head capsules, thoraxes, and elytra of selected prairie species yielded an age of 17,850 ± 550 yr B.P.; AMS dating of specimens of alpine tundra species yielded an age of 14,500 ± 500 yr B.P. (Elias and Nelson, 1989; Elias and Toolin, 1989). These first published AMS dates made directly from insect fossil exoskeletons indicate subtle reworking of the Lamb Spring sediments by spring activity, although this reworking was not observed in the stratigraphy.

Huntington Canyon, Utah (Fig. 7.6, No. 11), is a mammoth site with possible archaeological connections and is the highest-elevation mammoth site yet found in North America (2950 m above sea level). A projectile point (Pryor Stemmed Point) was found at the site in close proximity to a mammoth skeleton. As with the Lamb Spring site, there are dating problems. Spruce wood from below the mammoth skeleton was dated at 9700 yr B.P., but mammoth bone collagen has yielded an AMS date of 11,200 yr B.P. (Gillette and Madsen, 1992). I extracted insect fossils from the mammoth's skull cavity, and selected insect exoskeletons have yielded an AMS date of 10,500 yr B.P. (Elias, 1990b). The insects are indicative of a climate similar to modern conditions and reflect a shallow, sedge-lined pool in which the mammoth died.

Another set of archaeological sites originally believed to be of Pre-Clovis age are the Old Crow and Bonnet Plume Basin sites in the Yukon Territory (Fig. 7.6, Nos. 18 and 5). Initial excitement over very old bone "artifacts" led to extensive archaeological and paleontological investigations of these sites, including insect fossil analyses (Morlan and Matthews, 1978), although the insect fossils did not come from horizons containing artifacts. At present, most if not all of these bones are no longer considered bona fide artifacts, and other bones clearly modified by humans have been shown to be of Holocene age. These artifacts were not found in situ, but had weathered out of the bluffs at Old Crow. The age range of insects studied from Old Crow is from >40,000 to 30,000 yr B.P. (Matthews, 1975b; Matthews et al., 1990a; Morlan and Matthews, 1978, 1983). Regardless of archaeological problems, the Yukon sites in the Old Crow and Bonnet Plume Basins (Hughes et al., 1981) remain some of the most valuable paleontological sites in Eastern Beringia. These sites are discussed further in Chapter 10, dealing with Beringian paleoecology.

SOUTH AMERICAN ARCHAEOLOGY

At Monte Verde, Chile, an archaeological site dating to circa 13,000 yr B.P., described by Dillehay (1986, 1989), contains abundant wooden artifacts in peat; peat samples were taken by Ashworth et al. (1989) for fossil insect analysis. Samples from units MV-5 and MV-6 consisted of organic-rich sands and peats, coeval with the 13,000-yr B.P. artifacts. The identified beetle taxa represent two types of aquatic habitats: shallow, vegetation-rich water and riffles in a shallow stream. In addition, riparian beetles were found that represent open habitats. Rain forest–inhabiting species in the fauna are dominated by *Nothofagus* (southern beech)–associated taxa, including beetles that attack dead branches, decaying trunks, rotting wood, and leaf litter.

The environmental reconstruction based on the insect fauna is a shallow creek with sparsely vegetated sand bars and some boggy margins, flowing through rain forest. This reconstruction is consistent with that based on plant macrofossils, sediments, and diatoms (Dillehay, 1989). The late glacial paleoclimate was interpreted as very similar to the modern climate, an interpretation in agreement with all of the other lines of evidence except for the pollen interpretation. Heusser (1989) concluded from pollen studies that the climate was in transition from glacial to postglacial at 13,000 yr B.P., with substantially colder climates during the oldest human occupation (14,000–13,500 yr B.P.). In rebuttal, Ashworth et al. (1989) stated:

> The stratigraphic arrangement of samples in the base of MV-5, shown by Heusser, was interpreted to fit a preexisting regional pollen stratigraphy; only then do the cold-indicator species appear to be restricted to the basal portions of MV-5. Because of the broad temporal span (about 1500 years) represented by the indicator species, we do not believe that Heusser's argument for a climatic change is particularly convincing.

No synanthropic insect species were found in the Monte Verde assemblages. This is explained by Ashworth et al. (1989) through the following arguments:

1. European sites with synanthropes are all sites of accumulation of dung and rotting meat (e.g., wells, middens), whereas no such accumulations were found at Monte Verde.
2. Waterlogged meat scraps and dung in the bog would not attract these insects.
3. There may have been no species preadapted to synanthropy available in Chile in late glacial times.

Any or all of these factors may have played a role in excluding synanthropic insects from the site.

COMPARISON WITH EUROPEAN AND OTHER OLD WORLD STUDIES

Most New World studies are of sites that date from the late Pleistocene or earliest Holocene, whereas all but one of the Old World sites studied are mid- to late Holocene in age. Differences in types of human occupation and life-styles between these times are important. In Europe, the more or less sedentary life-styles from the Bronze Age through medieval times led to the accumulation of organic debris (in, for example, wells, cesspits, latrines, and trash heaps) and associated insect faunas. In contrast, Paleolithic hunter-gathers leading a transhumance life-style left very little organic debris on the landscape. Regrettably, the only substantial artifact records of these people are their stone tools. Their wanderings left little mark on the landscape, so that coeval insect faunas generally reflect undisturbed habitats.

Old World sites contain abundant synanthropic insect species; New World sites contain few, and most of these are human parasites in feces (most of which are not insects anyway). There appears to have been a lack of adaptation of

New World insects to synanthropic life-styles until much later in the Holocene. This situation was probably due to a lack of suitable human-induced habitats, at least in western North America, until late in prehistory. Very few Late Prehistoric and early historic archaeological sites in the New World have as yet been analyzed.

The lack of human influence on natural environments until the late Holocene has both advantages and disadvantages. It allows researchers studying most New World sites to examine natural biotic responses to macroclimate without the confounding effects of humanity. On the other hand, insects studied thus far from New World archaeological sites rarely shed much light on human living conditions and activities.

The field of Quaternary entomology has much to offer the archaeologist who is attempting to reconstruct both natural and anthropogenic environments. My aim is not to ridicule those archaeologists who have not yet availed themselves of the services of a paleoentomologist. Rather, I am anxious for collaborations to begin on a larger scale in more regions. Once this happens, we will gain new insights and spawn a new generation of questions important to both fields.

8

EUROPEAN STUDIES

The last word has not been said and much further work has still to be done.
—Van Geel et al. (1989)

The study of Quaternary insect fossils is most fully developed for sites in western Europe, especially Great Britain. These studies have made an impact on regional Quaternary studies, especially in the study of climate change. As elsewhere, most European research has focused fossil assemblages dating to the last glacial cycle. This chapter presents an outline of these studies.

BRITISH RESEARCH

The development of our knowledge of Quaternary insects owes the most to studies undertaken in the British Isles. More than fifty studies have been performed in this relatively small region (Table 8.1), exceeding the number of studies in either western Europe or Russia. Although the number of studies in North America is steadily increasing, the American research is spread out over a far greater area. The studied American fossil record contains many temporal and spatial gaps, whereas the British record provides data for many regions, covering faunas from the last interglacial onward. The only important gap in the British record is the latter half of the Holocene. Although there is no lack of fossil material, it is nearly impossible to discern whether these insect assemblages reflect natural or anthropogenic changes to the landscape during the last several thousand years.

I am sure that my British colleagues would argue, quite rightly, that there is still much to be done. However, because of its depth and clarity, the British record provides some unique insights, not only into the timing and intensity of regional environmental changes, but also into the responses of insect species and communities to those changes. When combined with paleobotanical and vertebrate paleontological studies, study of the British Quaternary insect record

Table 8.1

British Quaternary insect fossil sites and references

Site	Reference(s)
1. Abingdon	Aalto et al. (1984)
2. Aghnadarrah	McCabe et al. (1987)
3. Alcester	Shotton et al. (1977)
4. Austerfield	Gaunt et al. (1972)
5. Barnwell Station	Coope (1968a)
6. Beckford	Briggs et al. (1975a)
7. Bigholm Burn	Bishop and Coope (1977)
8. Bobbitshole	Coope (1974a)
9. Brandon	Coope (1968b)
10. Brighouse Bay	Bishop and Coope (1977)
11. Brimfield	Shotton et al. (1974)
12. Burnhead	Coope (1962a)
13. Chelford	Coope (1959)
14. Colnbrook	Coope (1982)
15. Colney Heath	Pearson (1967)
16. Corstorphine	Coope (1968c)
17. Croydon	Peake and Osborne (1971)
18. Dimlington	Penny et al. (1969)
19. Drumurcher	Coope et al. (1979)
20. Farmoor	Coope (1976)
21. Fisherwick	Osborne (1979)
22. Fladbury	Coope (1962b)
23. Folkestone	Coope (1980)
24. Four Ashes	Morgan (1973)
25. Glanllynnau	Coope and Brophy (1972)
26. Glen Ballyre	Coope (1971a)
27. Great Billing	Morgan (1969)
28. Hawkstor	Coope (1977)
29. Isleworth	Coope and Angus (1975)
30. Kempton Park	Gibbard et al. (1981)
31. Kirkby on Bain	Girling (1980)
32. Lea Marston	Osborne (1973)
33. Lea Valley	Coope and Tallon (1983)
34. Low Wray Bay	Coope (1977)
35. Marsworth	Green et al. (1984)
36. Nechells	Shotton and Osborne (1965)
37. Northmoor	Coope (1976)
38. Porthmeare	Osborne (1976)
39. Queensford	Briggs et al. (1985)
40. Redkirk Point	Bishop and Coope (1977)
41. Red Moss	Ashworth (1972)

Table 8.1 (*Continued*)

Site	Reference(s)
42. Roberthill	Bishop and Coope (1977)
43. Rodbaston	Ashworth (1973)
44. Shortalstown	Coope (1971b)
45. Shustoke	Kelly and Osborne (1965)
46. St. Aubins Bay	Coope et al. (1985)
47. St. Bees	Coope and Joachim (1980)
48. Stafford	Morgan (1970)
49. Sugworth	Briggs et al. (1975b), Osborne (1980a)
50. Syston	Bell et al. (1972)
51. Tame Valley	Coope and Sands (1966)
52. Tattershall	Girling (1974, 1980)
53. Trafalgar Square	Franks et al. (1958)
54. Upton Warren	Coope et al. (1961)
55. West Bromwich	Osborne (1980b)
56. West Drayton	Coope (1982)
57. Wilden	Shotton and Coope (1983)
58. Wilsford	Osborne (1969)
59. Wretton	Coope (1974b)

also reveals the different ways in which these ecosystem components have responded to change.

As Coope (1988) pointed out, during the late Quaternary the British Isles have been on the receiving end of some of the greatest fluctuations in sea surface temperatures anywhere in the world. This is because, at various times, Great Britain has either basked in the warmth of the Gulf Stream (as it does today) or been chilled by the icy waters of the North Atlantic polar front (see Ruddiman and McIntyre, 1981). The effects of ocean currents, combined with Milankovitch cycles in solar radiation through the late Quaternary, have created climatic episodes in Britain varying from Mediterranean warmth to Tibetan cold (Coope, 1987b).

In Britain, as elsewhere, it is very difficult to assign a reliable age to most fossil assemblages older than the last interglacial. Many events that occurred within the last 40,000 yr can be dated by radiocarbon (given suitable quantity and quality of organic preservation), but beyond this interval are hundreds of thousands of years that fall within the gap between radiocarbon and other isotopic dating methods. Some sediments and fossils in this age range have been successfully dated through thermoluminescence dating (Forman, 1989), amino acid racemization of proteins in shells and bone (Miller and Brigham-Grette, 1989; Geyh and Schleicher, 1990), and tephrochronology (Walter,

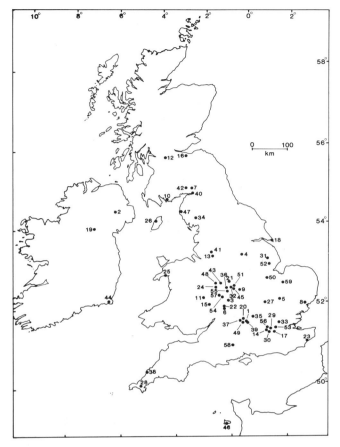

Figure 8.1. Map of the British Isles showing Quaternary insect fossil localities. Site numbers are keyed to the list of sites in Table 8.1.

1989). However, many deposits laid down before the last interglacial have not been dated accurately. For instance, Briggs et al. (1975b) described an inter-glacial deposit containing insect fossils, plant macrofossils, pollen, mollusks, and vertebrates from Sugworth, Great Britain (Fig. 8.1, No. 49). Based on stratigraphic position and correlation with other floras and faunas, this site was placed in the Cromerian Interglacial (two or possibly three interglacials prior to the last interglacial). But, because there is no radiometric dating of the site, this remains a tentative assignment.

With these problems in mind, I focus most of this chapter on fossil insect faunas that date from the last interglacial through the Holocene. Even if reliable dates on older assemblages could be obtained, the number of these that have

Table 8.2

Correlation of British and Continental European Pleistocene sequences with oxygen isotope stages

Isotope stage[a]	Approximate age (yr B.P.)[b]	British sequence[c]	Continental sequence[d]
1	0–10K	Flandrian	Holocene
2	13–26K	Late Devensian Stadial	Late Würm/Weichselian Stadial
3	26–50K	Upton Warren Interstadial Complex	Denekamp/Hengelo Interstadial
4	50–79K	Mid-Devensian Stadial	Glinde/Oerel Stadial?
5a	79–112K	Brimpton Interstadial	St. Germain II/Odderade
5c		Chelford Interstadial	St. Germain I/Brörup
5e	122–132K	Ipswichian Interglacial	Eemian Interglacial
6	132–198K	Wolstonian Stadial	Saale Glaciation

[a] Isotope stages after Shackleton and Opdyke (1973).
[b] Ages after Bowen et al. (1986).
[c] British correlation with oxygen isotope stages after Ehlers et al. (1991).

been studied is quite small in proportion to the number of younger assemblages. Yet we have a considerable amount of information on last-interglacial insect faunas. In Britain, this interglacial is called the Ipswichian, after the type locality at Ipswich, Suffolk (Table 8.2). During the Ipswichian, Britain was home to stenothermic ground beetle species that today live in southern Europe. This finding indicates that the climate at the height of the last interglacial (isotope stage 5e) was warmer than the modern climate (Coope, 1990). Other lines of evidence point in the same direction. For instance, the members of the thermophilous insect fauna recorded from Ipswichian swamp deposits at Trafalgar Square, London (Fig. 8.1, No. 53) were coinhabitants with the hippopotamus (Franks et al., 1958). Steppe indicators, such as horses, are absent from the Ipswichian faunas (Moore, 1986).

The fossil record shows that the Ipswichian biota enjoyed a relatively long, stable climatic regime that was sufficiently protracted to allow establishment of mature deciduous forests, including birch and maple, accompanied by their particular bark beetles (Coope, 1990). Along with the subtropical mammalian megafauna, a rich dung beetle fauna also occupied Britain during this interglacial. Several species were found that today inhabit only the Mediterranean region.

Coope (1974a) described a fossil insect assemblage from the type locality at Bobbitshole, Ipswich (Fig. 8.1, No. 8). The fauna was deposited in a shallow, eutrophic pond or lake with abundant marsh vegetation. The modern distribution of stenotherms in the assemblage indicates that summer temperatures were about 3°C warmer than at present. The composition of the assemblage closely matches that from Trafalgar Square.

The Ipswichian Interglacial ended after about 122,000 yr B.P., as recorded in fossil assemblages that show a replacement of thermophilous beetles with arctic and subarctic species. The fauna that included Mediterranean species was replaced by a fauna resembling that of modern-day Scandinavia and Siberia. Again, the exact timing of these events remains uncertain. The next warm interval, the Chelford Interstadial in Britain (Table 8.2), took place sometime early within the last (Devensian) glaciation. Coope (1977) placed the Chelford Interstadial between about 64,000 and 54,000 yr B.P. Subsequently Bowen et al. (1986) placed the Chelford before 112,000 yr B.P. Insect faunas from this interstadial, including assemblages from Chelford, Four Ashes, and Wretton (Fig. 8.1, Nos. 13, 24, and 59), are composed of species with modern distributions overlapping in southern Finland (between 60° and 65° N latitude). The Chelford Interstadial is thus interpreted as being warmer than prior and subsequent stadial episodes, but with summer temperatures perhaps 2–3°C cooler than modern ones (Coope, 1977). McCabe et al. (1987) described an interstadial flora and fauna from Aghnadarrah, Northern Ireland (Fig. 8.1, No. 2) that is probably also of Chelford age.

Next comes the Upton Warren interstadial complex in Britain (Table 8.2), named for deposits at the Upton Warren site (Fig. 8.1, No. 54) near Birmingham. The timing of this interstadial is less in question because it is within the range of radiocarbon dating. Coope (1975, 1977) summarized the Upton Warren faunas, dating them between 45,000 and 25,000 yr B.P. (they were more recently redated by Bowen et al., 1986, from 50,000 to 26,000 yr B.P.). The Upton Warren interval was a complex of environmental changes that began with a rather abrupt warming that lasted roughly 5000–6000 yr, followed by 15,000–16,000 yr of conditions cooler than modern, but not fully glacial. The climatic deterioration into the late Devensian Stadial occurred at about 24,000 yr B.P. (Fig. 8.2).

A boreo-montane fauna dominated British assemblages just prior to the rapid warming in the Upton Warren. The climate was arctic in character, though perhaps not continental, given the lack of Asian beetles (in contrast to the subsequent late Devensian Stadial faunas). However, the shift to warm conditions was apparently quite rapid, as evidenced in the Tattershall, Lincolnshire assemblages (Fig. 8.1, No. 52) (Girling, 1974, 1980), in which the

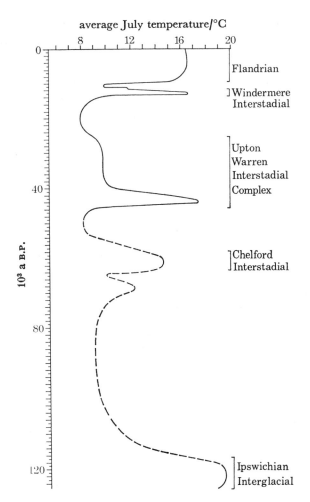

Figure 8.2. Reconstruction of mean July temperatures in Britain, based on insect fossil assemblages. (From Coope, 1977.)

arctic fauna was replaced by a temperate fauna within a few centimeters of sediment. The markedly thermophilous character of the 43,000-yr B.P. fauna from Isleworth, near London (Fig. 8.1, No. 29), also supports the rapidity of this change (Coope and Angus, 1975). The Isleworth fauna documents the brief rise in summer temperatures to levels above modern. Yet throughout this period there is no evidence of trees occurring in Britain (Coope et al., 1961). This fauna is extremely rich, comprising 248 identified species in 34 families. The diversity of the fauna shows the degree of development of temperate insect

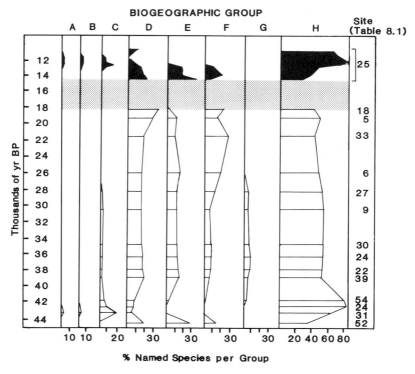

Figure 8.3. Percent composition of British insect fossil assemblages during the middle and late Devensian glaciations, based on modern geographic distribution of species. Stippled zone represents interval devoid of insect fossils. Columns: A, southern European species; B, southern European species whose ranges just fail to reach Britain; C, southern species whose ranges are south of central Britain; D, widespread species whose ranges are north of central Britain; E, boreal and montane species whose ranges extend down into the upper part of the coniferous forest zone; F, boreal and montane species whose ranges are above treeline; G, eastern Asiatic species, some of which range into North America; H, cosmopolitan species (wide geographic ranges). Site numbers are keyed to the list of sites in Table 8.1. (Modified from Coope, 1987a.)

communities, even in times of rapidly changing climate. This level of diversity is especially striking given that the climatic optimum of the interstadial may have persisted only 1000 years (Coope, 1987a,b).

The climatic deterioration that followed was much more gradual. Several assemblages from the Four Ashes site in the Midlands (Fig. 8.1, No. 24) (Morgan, 1973) reflect these conditions. They contain a mixture of temperate and northern species and lack the extreme thermophiles seen from Isleworth and Tattershall. Later in the interstadial complex, further deterioration is marked in assemblages from the Upton Warren site (Coope, 1959). These assemblages are also noteworthy because they include a considerable number of Asiatic species, such as the Tibetan dung beetle, *Aphodius holdereri,* suggesting a shift to continental climate.

Coope (1987b) summarized the insect faunal sequences between 25,000 and 10,000 yr B.P. (Fig. 8.3). As ice sheets approached the English Midlands from the north, the remaining insect faunas south of the ice became increasingly impoverished, with just a few species identified from deposits dating to the late Devensian Stadial maximum, at about 20,000 yr B.P. (Table 8.2). Between 18,000 and 15,000 yr B.P., even southern Britain experienced extremely cold, harsh climates, with widespread ice wedge formation in the lowlands and no evidence of any plants or animals in the pond and lake deposits of this period. Conditions were probably similar to those of a modern polar desert. Coope (1987b) speculated that the beetle fauna may have been extirpated from Britain at this time.

I discussed the British late glacial insect faunas in some detail in Chapter 5. As in the Upton Warren Interstadial Complex, the late glacial insect faunal sequence shows remarkably rapid and dramatic changes from arctic faunas prior to 13,000 yr B.P. at Hawkstor, Cornwall (Fig. 8.1, No. 28) (Coope, 1977) and Glanllynnau, northern Wales (Fig. 8.1, No. 25) (Coope and Brophy, 1972), to temperate faunas during the brief Windermere Interstadial. Insects from a deposit at Low Wray Bay, Windermere (Fig. 8.1, No. 34) document the rapid warming following 13,000 yr B.P. (Coope, 1977), as do insects from the Roberthill site in southern Scotland (Fig. 8.1, No. 42) (Bishop and Coope, 1977). Again, the Windermere Interstadial faunas reflect summer conditions as warm as modern ones. Insect faunas dating from between 12,200 and 11,000 yr B.P. from a number of sites suggest a gradually cooling, but oscillating, climate.

The Windermere Interstadial was followed by a striking deterioration between 11,000 and 10,000 yr B.P. This is called the Loch Lomond Stadial in Britain. It correlates with the Younger Dryas pollen zone in continental Europe. Cool temperate faunas were replaced once more with arctic species, including the return of several Asian species (Coope, 1977). The subsequent amelioration to the Flandrian, or Holocene, was equally rapid and extreme, bringing summer temperatures as warm as or warmer than modern ones by 9500 yr B.P. (Coope, 1987b). Most of the British Holocene insect assemblages that have been studied come from archaeological sites (Chapter 7).

RESEARCH IN THE NORTH ATLANTIC REGION

Fossil arthropod research is now underway in Iceland and Greenland. Haarløv (1967) published the first results of studies of Quaternary insect fossils from Greenland. He studied late Holocene peats from Sermermiut, near Jacob-

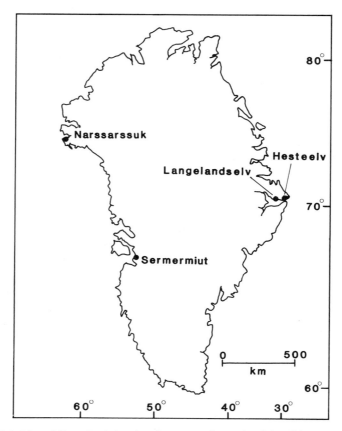

Figure 8.4. Map of Greenland showing Quaternary insect fossil localities.

shaven, western Greenland (Fig. 8.4). The peats contained oribatid mites and the remains of leaf miner flies (Diptera: Agromyzidae). The fauna comprises species associated with bogs and humid heaths in arctic and subarctic regions, though three of the mites identified from the assemblages have not been collected in modern Greenland.

Böcher and Bennike (1991) described Eemian interglacial (isotope stage 5e) insect assemblages from Hesteelv and Langelandselv, Jameson Land, east Greenland (Fig. 8.4). The assemblages contained only eight insect taxa, but the modern ranges of the identified beetle species are considerably south of the fossil site. Böcher and Bennike concluded that the fossil assemblages reflect climatic conditions markedly warmer than those of today.

Böcher (1989b) also reported on the discovery of *Amara alpina* (Carabidae) fossils in an organic deposit at Narssarssuk in northwestern Greenland

Figure 8.5. Map of Europe showing Quaternary insect fossil localities. Site numbers are keyed to the list of sites in Table 8.3.

(Fig. 8.4). The deposit contained mollusk shells that yielded amino acid racemization ratios equivalent to those found in regional shells tentatively assigned to isotope stage 5a. In Greenland, this interval is called the Qarmat Interstadial. *A. alpina* is absent from Greenland today, and the fossil locality is also slightly farther north than the farthest northern populations of that species in nearby arctic Canada.

Buckland et al. (1986a,b) have investigated a number of sites in Iceland, comparing insect faunas before and after the arrival of Viking settlers. The Viking landings conveniently coincided with the deposition of the Landnám tephra in Iceland. The tephra thus serves as a marker horizon, separating the assemblages into pre- and postsettlement faunas. I have previously discussed the post-Landnám (archaeological) faunas in Chapter 7. Pre-Landnám faunas have been described by Buckland et al. (1986b) from Einhyrningur, Hofsá, Holt, Kópavogur, Merkigil, Ósabakki, Skálafelsjökull, and Thjórsárbrú (Fig. 8.5, Nos. 9, 15, 16, 18, 31, 36, 42, and 44) and from Ketilsstadir (Fig. 8.5, No. 17) (Buckland et al., 1986a). The pre-Landnám fauna was considerably less diverse than the post-Landnám because of the successful establishment of Viking-

Table 8.3
Continental European Quaternary insect fossil sites and references

Site	Reference(s)
1. Andøya, Norway	Fjellberg (1978)
2. Artuki, Byelorussia	Nazarov (1984 and written communication, 1991)
3. Bad Tatzmannsdorf, Austria	Schweiger (1967)
4. Belchatów, Poland	Morgan et al. (1982)
5. Björkeröds Mosse, Sweden	Lemdahl (1985, 1988b)
6. Bollnäs, Sweden	Lindroth (1948)
7. Brumunddal, Norway	Helle et al. (1981)
8. Cotentin Peninsula, France	Coope et al. (1987)
9. Einhyrningur, Iceland	Buckland et al. (1986b)
10. Grossensee, Germany	Günther (1983)
11. Håkulls Mosse, Sweden	Berglund et al. (1984), Lemdahl (1985, 1988b)
12. Härnön, Sweden	Lindroth (1948)
13. Le Havre, France	Ters et al. (1971)
14. High Ardennes, France	Damblon et al. (1977)
15. Hofsá, Iceland	Buckland et al. (1986b)
16. Holt, Iceland	Buckland et al. (1986b)
17. Ketilsstadir, Iceland	Buckland et al. (1986a)
18. Kópavogur, Iceland	Buckland et al. (1986b)
19. La Taphanel, France	Ponel and Coope (1990)
20. Lac d'Issarlés, France	Ponel and Gadbin (1988)
21. Lago di Monterosi, Italy	Roback (1970)
22. Lake Balaton, Hungary	Dévai and Moldován (1983)
23. Lake Bysjön, Sweden	Lemdahl (1988c)
24. Längsele, Sweden	Lindroth (1948)
25. Lausanne, Switzerland	Gabus et al. (1987)
26. Leveäniemi, Sweden	Lindroth and Coope (1971)
27. Lobsigensee, Switzerland	Elias and Wilkinson (1983)
28. Logoza, Byelorussia	Nazarov (written communication, 1991)
29. Maar lakes (multiple sites), Germany	Hofmann (1990)
30. Mark Valley, The Netherlands	Bohncke et al. (1987)
31. Merkigil, Iceland	Buckland et al. (1986b)
32. Mickelsmossen, Sweden	Lemdahl and Persson (1989)
33. Niederwenigen, Switzerland	Elias and Schlüchter (1993)
34. Nikol'skoye, Russia	Bidashko and Proskurin (1988)
35. Nizhninsky Rov Ravine, Byelorussia	Nazarov (written communication, 1991)
36. Ósabakki, Iceland	Buckland et al. (1986b)
37. Peelo, The Netherlands	Coope (1969)

Table 8.3 (*Continued*)

Site	Reference(s)
38. Piilonsuo, Finland	Koponen and Nuorteva (1973), Karppinen and Koponen (1974)
39. Pilgrimstad, Sweden	Lindroth (1948)
40. Poolsee, Germany	Hofmann (1983)
41. Rubezhnitsa, Byelorussia	Nazarov (1979)
42. Skálafelsjökull, Iceland	Buckland et al. (1986b)
43. Taillefer Massif, France	Ponel et al. (1992)
44. Thjórsarbrú, Iceland	Buckland et al. (1986b)
45. Timoshkovichi, Byelorussia	Nazarov (written communication, 1991)
46. Toppeladugård, Sweden	Lemdahl (1985, 1991a)
47. Torreberga, Sweden	Berglund and Digerfeldt (1970)
48. Usselo, The Netherlands	Van Geel et al. (1989)
49. Voorthuizen, The Netherlands	Angus (1975)
50. Zabinko, Poland	Lemdahl (1991b)

introduced species. The pre-Landnám faunas, all of Holocene age, are adapted to cool-temperate rather than arctic climates. Coope (1969, 1986) and Buckland et al. (1986b) have argued that the Icelandic insect fauna did not survive the Weichselian glaciation in situ, but rather migrated to Iceland following deglaciation. Fully winged species may have been carried to Iceland from northwest Europe by winds, but many of the pre-Landnám beetle species are flightless. They proposed a transport mechanism for these beetles in which sediment-covered ice floes from Scandinavia and northern Britain drifted to the northwest to Iceland, beaching their passengers on Iceland's newly deglaciated shores. This intriguing hypothesis is in need of additional tests.

WESTERN EUROPEAN RESEARCH

Although the number of studies of insect faunas from continental Europe is substantially smaller than that of faunas from Britain, good progress has been made in many regions during the last twenty years (Table 8.3).

This is especially true for Scandinavia, largely through the efforts of Geoffrey Lemdahl at the University of Lund, Sweden. As discussed in the first chapter, Scandinavian paleoentomological research began in earnest with Carl Lindroth's (1948) work on interglacial insect fossils from Bollnäs, Härnön, Läng-

Table 8.4

Correlation of Late Glacial chronozones in Britain and continental Europe

Radiocarbon age (yr B.P.)[a]	British zone	Continental zone
10,000–Recent	Flandrian	Holocene
10,000–11,000	Loch Lomond Stadial	Younger Dryas Chronozone
11,000–12,000	Windermere Interstadial	Alleröd Chronozone
12,000–13,000	Windermere Interstadial	Bölling Chronozone
13,000–>14,500	Late Devensian Stadial	Oldest Dryas Chronozone

[a]Late glacial chronology after Lotter (1991).

sele, and Pilgrimstad in central Sweden (Fig. 8.5, Nos. 6, 12, 24, and 39). Later, Lindroth and Coope (1971) reported their conclusions about insect fossils from deposits thought to be of interglacial age at Leveäniemi (Fig. 8.5, No. 26). The composition of the Leveäniemi assemblages was different from that of the other "interglacial" faunas in that the Leveäniemi assemblages contain substantially more thermophiles. Even the Längsele fauna, the most thermophilous of the interstadial faunas, reflects conditions colder than those at present. The Leveäniemi fauna is indicative of climatic conditions considerably warmer than those at present and suggests a continental climate.

Based on these differences and stratigraphic evidence developed by Lundqvist (1967), Lindroth and Coope ascribed the Bollnäs, Härnön, Längsele, and Pilgrimstad faunas to an early Weichselian Interstadial, possibly correlative with the Brörup of Denmark. The Brörup has been radiocarbon-dated at 59,430 yr B.P. (Lundqvist, 1986), but radiocarbon ages older than about 45,000 yr B.P. should be treated as minimum ages (see Table 8.2 for correlations and chronology). This scheme has been widely accepted in subsequent stratigraphic and paleoenvironmental studies in Sweden (Robertsson, 1988). However, Lundqvist (1986) has assigned these interstadial sites to a Jämtland Interstadial that is slightly younger than the Danish Brörup.

Scandinavia was covered by ice during the late Weichselian Glaciation, which began about 47,000 yr B.P. Southern Sweden was deglaciated by about 14,000 yr B.P., and northern Sweden by about 8500 yr B.P. (Lundqvist, 1986). Organic deposits in southern Sweden contain insect fossils documenting late glacial conditions, including an initial warming, followed by the Younger Dryas oscillation (see Table 8.4 for late glacial chronozones and correlations).

Lemdahl has examined a number of late glacial sites in Scania, southern Sweden. Insect faunas from the Håkulls Mosse site (Fig. 8.5, No. 11) span the interval 13,000–10,000 yr B.P. Prior to 12,500 yr B.P., southern Sweden was a periglacial landscape with stagnant ice, sparse vegetation cover, and soils made unstable by solifluction (Berglund et al., 1984). The earliest deposit containing insects dates from 13,000 yr B.P. This assemblage contains thermophilous insects that live today in the boreal and subalpine zones, but the vegetation at that time (Bölling pollen zone) was composed of plant species that today are members of steppe-tundra communities. The insect assemblages continue to show warm conditions until about 12,500 yr B.P., when a mixed fauna of temperate and northern species indicates a gradual climatic deterioration during the Alleröd pollen zone. Boreal and subalpine vegetation failed to become established in southern Sweden during the relatively brief late glacial interstadial. In the insect record, the interstadial fauna was replaced by arctic and alpine species at about 11,000 yr B.P., signaling the Younger Dryas oscillation. Climatic amelioration was documented in the insect record at Håkulls Mosse from 10,500 yr B.P. onward, but the vegetation record lagged behind this change by about 300 years.

Additional work on late glacial insect faunas from the Björkeröds Mosse and Toppeladugård sites in Scania (Fig. 8.5, Nos. 5 and 46) (Lemdahl, 1985) and Alleröd faunas from Lake Bysjön (Fig. 8.5, No. 23) (Lemdahl, 1988c) helped to refine the regional paleoenvironmental reconstruction. Again, the insect interpretation was considerably different from traditional paleobotanical reconstructions (Lemdahl, 1985). The insect records show that a brief, sudden amelioration began in Scania about 12,600 yr B.P., followed by gradual cooling that began by 12,300 yr B.P., culminating in Younger Dryas cooling following 11,300 yr B.P. that saw the return of arctic conditions and fauna. The additional sites confirmed the Håkulls Mosse data showing rapid amelioration beginning at 10,500 yr B.P.

Lemdahl (1991a) applied the mutual climatic range (MCR) method to the Toppeladugård assemblages to reconstruct summer and winter mean temperatures from 11,800 to 10,000 yr B.P. The MCR reconstruction shows mean summer temperatures of 15–18°C from 11,800 to 11,600 yr B.P. and mean summer temperatures of 10–13°C by 10,800 yr B.P. Following the Younger Dryas, summer temperatures rose to 14–16°C by 10,200 yr B.P.

Only two insect fossil studies have been reported from Norway. The first was by Coope (in Helle et al., 1981). This study concerned early Weichselian Interstadial deposits at Brumunddal, southeastern Norway (Fig. 8.5, No. 7). Both pollen and insect data indicate climatic conditions with summer temperatures 2–3°C cooler than those at present. The lower of two insect assemblages is comprised of species found today in central Scandinavia (i.e., slightly north of the fossil locality). The upper assemblage is distinctly different; it contains only cold-adapted, arctic species. Taken together, the upper fauna indicates summer temperatures at or just below 10°C. Coope correlated the Brumunddal insect faunal sequence with the Jämtland interstadial faunas of central Sweden. Lundqvist (1986) correlated the Brumunddal deposits with the Brörup interstadial of Denmark.

The other Norwegian study was a brief treatment by Fjellberg (1978) of mid- and late Weichselian insects from lake sediments at Andøya, northern Norway (Fig. 8.5, No. 1). The lake sediments show continuous deposition from >18,000 yr B.P. to the present. The record from 18,000 to 12,000 yr B.P. shows arctic tundra vegetation and arctic chironomids and oribatid mites.

The only significant paper on Quaternary insects of Finland is the study by Koponen and Nuorteva (1973) on the peat bog of Piilonsuo (Fig. 8.5, No. 38). The faunal assemblages range in age from 10,000 to 600 yr B.P. and include abundant conifer-associated species (many specimens were taken from under the preserved bark of buried tree stumps). In total, 259 taxa from more than 70 families in eleven orders of insects and arachnids were recovered. All of the identified species occur in Finland today. The fossil mite faunas were reported separately by Karppinen and Koponen (1974). The faunal assemblages provide a wealth of paleoecological data, reconstructing the history of forest and bog development in southern Finland.

In Germany, most published studies to date have been paleolimnological studies of lake sediments, including chironomid larvae. Hofmann (1983) studied chironomids from late glacial through Holocene-age sediments from Poolsee in northern Germany (Fig. 8.5, No. 40). The late glacial chironomids were all cold stenotherms. Postglacial faunas reflected increasing temperatures and changes brought on by siltation of the lake.

Hofmann (1990) also studied chironomids from the maar lakes (low-relief lakes formed by multiple, shallow, explosive volcanic eruptions, surrounded by a crater ring) on the Eifel Plateau in western Germany (Fig. 8.5, No. 29). The midge larval faunas range in age from the Eemian Interglacial through late Weichselian (16,000 yr B.P.). The late Weichselian sediments yielded few fossils. An interstadial fauna from sediments directly beneath the late Weichselian zone was more diverse, but the chironomid fauna still reflected cold climatic conditions. The Eemian fauna was diverse and abundant, reflecting the warmth of the interglacial.

Günther (1983) investigated chironomid larvae from Grossensee lake sediments in the Holstein region of Germany (Fig. 8.5, No. 10). The sediments span the Holocene. The chironomids, along with other fossil invertebrates (Cladocera and ostracods), document anthropogenic changes in the trophic status of the lake.

British workers have undertaken some fossil beetle research from late Pleistocene sites in the Netherlands. Angus (1975) studied early Weichselian stadial faunas (stratigraphically positioned between the Brörup and Hengelo interstadials) from Voorthuizen (Fig. 8.5, No. 49). A lower fossil bed yielded an arctic fauna, reflecting summer temperatures below 10°C. A younger assemblage was less climatically diagnostic but was thought to indicate a slight warming.

Coope (1969) studied beetles from Peelo, from a site that may be of mid-Weichselian age (Fig. 8.5, No. 37). This fauna was indicative of glacial stadial conditions and included the tundra beetle *Pterostichus vermiculosus,* which lives today only in arctic regions of North America and Siberia.

Coope (in Bohncke et al., 1987) also studied late glacial insect assemblages from deposits at Notsel in the Mark Valley of the Netherlands (Fig. 8.5, No. 30). The assemblages range in age from about 12,600 to 10,300 yr B.P. The insect faunas from a peat unit (12,600–10,970 yr B.P.) showed little change. Using the MCR method, Coope estimated mean summer temperatures of 15–18°C for this interval. The beetle fauna from an overlying sand unit (10,900–10,300 yr B.P.) documents the marked cooling associated with the Younger Dryas chronozone. The temperate fauna of the peat unit was replaced by a fauna of arctic affinities. MCR estimates of summer temperatures associated with this fauna were 10–11°C.

Van Geel and Coope (in Van Geel et al., 1989) examined late glacial mites and insects from Usselo, the Netherlands (Fig. 8.5, No. 48). Beginning at 13,000 yr B.P., the arthropod data suggest warm conditions (mean summer temperatures of 15–20°C). A slight cooling was recorded by the fauna of the Alleröd zone, followed by substantial cooling during the Younger Dryas (mean summer temperatures 10–11°C). As in British and Swedish late glacial studies, the arthropod results from Usselo differed somewhat from paleobotanical interpretations.

More extensive research has been carried out in France, by both British and French researchers. On the Cotentin Peninsula (Fig. 8.5, No. 8), Coope et al. (1987) studied insect assemblages from coastal deposits of Eemian to early Weichselian age. The oldest studied faunas reflect interglacial conditions corresponding in the paleobotanical record with cool-temperate mixed forest. Above this unit, the flora and insect fauna are indicative of arctic-subarctic conditions. Finite radiocarbon ages of various regional deposits were considered to be far too young. Based on stratigraphic correlation with uranium series–dated sites from nearby Jersey, the cold-adapted fauna may be 115,000 yr old, signaling rapid climatic deterioration following the Eemian Interglacial.

Ponel and Coope (1990) have completed an important study of late glacial and early Holocene insect faunas in lake sediments from La Taphanel in the Massif Central region (Fig. 8.5, No. 19). The paleoclimatic sequence from La Taphanel was described in Chapter 5. The beetle fauna documents a rapid climatic warming beginning at 13,000 yr B.P. The paleobotanical signal was substantially out of synchroneity with the insect signal until the Alleröd zone.

Unlike assemblages from elsewhere in Europe, the Taphanel faunas show an episode of cooling between temperate Bölling zone assemblages and cooler Alleröd faunas. This cooling corresponds to the Older Dryas pollen zone. Marked cooling occurred in the Younger Dryas, followed by rapid amelioration after 10,300 yr B.P.

Studies of Holocene environments in France include insect fossil analyses from peat deposits in the High Ardennes region of northeast France (Fig. 8.5, No. 14) (Damblon et al., 1977) and Ponel and Gadbin's (1988) study of lake sediments from Lac d'Issarlés in the Ardèche region of southeastern France (Fig. 8.5, No. 20). The Ardennes study chronicles the effects of human activity on the ecology of regional peat bogs, and the Lac d'Issarlés study verified paleobotanical reconstructions of the regional establishment of hazel (*Corylus avellana*) forests.

Ponel et al. (1992) have worked on insect fossil assemblages from the Taillefer Massif region of the French Alps (Fig. 8.5, No. 43). They reconstructed treeline fluctuations for the Holocene and documented the faunal and floral succession from open ground tundra through mature forest. After 2000 yr B.P., the forest shifted downslope. The beetle evidence helped to establish that the cause of this movement was climatic cooling rather than human influence.

Finally, Shotton and Osborne (in Ters et al., 1971) studied fossil insects from early Holocene estuarine deposits at Le Havre (Fig. 8.5, No. 13). The lower deposit (dated 8470 yr B.P.) contained a fauna associated with a reed swamp environment. Middle and upper deposits (older than 7820 yr B.P.) yielded more diverse faunas, indicative of a variety of aquatic and semiaquatic habitats. Although marshland deposits such as these might be taken to represent salt marsh environments, the beetle fauna lacks halobionts (salt-tolerant species), and Shotton and Osborne argue that the marshes grew in fresh water. All of the identified beetle species live in northern France today, so the climate was probably very similar to modern parameters.

The oldest Quaternary fossil insect material thus far studied from Switzerland comes from the Niederwenigen site, near Zurich (Fig. 8.5, No. 33) (Elias and

Schlüchter, unpublished data). The fossils are from early Weichselian–age sediments. The assemblages include species with modern distributions limited to arctic and alpine regions of Scandinavia, Siberia, and the Altai Mountains of Central Asia. The local environment was an open ground tundra with alder, birch, and willow shrubs. Paleoclimate reconstructions based on insect assemblages suggest average summer temperatures ranging from 15°C for the basal sample to 10°C for the uppermost sample. Mean January temperature estimates range from –10°C for the basal assemblage to –14.5°C for the youngest assemblage. Precise dates are unavailable for these faunas, but they appear to document the transition from an early Weichselian Interstadial to a stadial.

Studies on late glacial insect faunas in Switzerland began with the investigation of Lobsigensee, on the Swiss Plateau (Fig. 8.5, No. 27) (Elias and Wilkinson, 1983), in which a group of cold-adapted insects at the base of the sequence was rapidly replaced by a thermophilous group. The date of the climatic amelioration was about 12,900 yr B.P., but it should be emphasized that this is an interpolated age, not one fixed by radiocarbon dates at the climatic boundary. Based on beetle and caddisfly data, estimated mean July temperatures at the site remained at 14–16°C through the Bölling and Alleröd intervals. Insect preservation in Younger Dryas–age sediments was poor, and inferences drawn from insects in this interval are inconclusive and offer little positive evidence of a Younger Dryas cooling.

As mentioned in Chapter 7, Russell Coope and I investigated a second late glacial site in Switzerland, the Champreveyres site at Neuchâtel (Fig. 7.3, No. 3). Although this is an archaeological site, the late glacial insect faunas reflect essentially nonanthropogenic environments, and as such provide useful paleoenvironmental evidence. The Champreveyres insect data show the intensity of the late glacial climatic amelioration. This site contains stenothermic thermophiles (e.g., the ground beetle *Calosoma inquisitor*) very close to the beginning of the late glacial period. The degree of increase in summer temperatures (about 7°C) was of the same order in northwestern Europe. This sudden and intense warming episode was synchronous over much of western Europe. Following the initial warming, an ecologically diverse beetle fauna became established at Champreveyres. Unfortunately, the Younger Dryas zone is represented by a coarse gravel deposit with little organic preservation, so the Champreveyres insect sequence is uninformative about this period.

Lemdahl (in Gabus et al., 1987) described a late glacial insect assemblage from a terrace above Lake Geneva at Lausanne (Fig. 8.5, No. 25). The terrace deposits date to 13,210 yr B.P. The insect fauna is indicative of open ground environments but lacks arctic and alpine species. It is essentially a temperate zone fauna, in contrast to the fossil flora, which suggests arctic tundra.

In eastern Austria, Schweiger (1967) investigated late glacial insects from a borehole at Bad Tatzmannsdorf (Fig. 8.5, No. 3). Although the samples contained mixed horizons, Schweiger discerned two distinct faunas. The majority of specimens were thought to be associated with the Alleröd pollen zone. This suite comprises thermophilous beetles that are found today in southern Europe, including arid regions of the eastern Mediterranean. The other faunal group, which includes boreo-montane beetles, has been assigned to an earlier, cold climatic interval, which was correlated with the Older Dryas pollen zone based on the palynology of the sediments.

Roback (1970) studied chironomid larvae as part of an interdisciplinary limnological study of Lago di Monterosi in southern Italy (Fig. 8.5, No. 21). Lake sediments have accumulated in this basin for more than 23,000 years. Prior to the late Weichselian (i.e., older than 23,000 yr B.P.), the chironomids reflect a shallow, slightly alkaline lake. From 23,000 to 5,000 yr B.P., the chironomid fauna was absent or in a poor state of preservation. During the lengthy hiatus, the lake appears to have been too shallow and too low in organic nutrients to support chironomids and most other invertebrates. The mid-Holocene fauna was the most abundant and diverse, probably in response to rising lake levels and increasing organic content. From that time on, the chironomid fauna showed periodic cycles of abundance and paucity, mirroring fluctuations in trophic status and water levels in the lake.

EASTERN EUROPEAN STUDIES

Outside Russia, only one published study in Quaternary entomology has been carried out by eastern European scientists: a paleolimnological study of Lake Balaton, Hungary (Fig. 8.5, No. 22), by Dévai and Moldován (1983). Sediments in Lake Balaton span the last 20,000 years. The authors studied fossil chironomid larvae from shallow, undated sediment cores and traced the history of lake eutrophication over the last few centuries. They noted a sharp increase in eutrophication during the last few decades.

Polish research includes just two studies of late Weichselian sites. Morgan et al. (1982) published an abstract about a small, mid-Weichselian fauna from Belchatów in southeastern Poland (Fig. 8.5, No. 4). The fauna, including *Holoboreaphilus nordenskioeldi* and *Diacheila polita,* is indicative of open ground conditions under a cold, continental climate, prior to 25,000 yr B.P.

Lemdahl (1991b) published the results of a study on late glacial insects from Zabinko in western Poland (Fig. 8.5, No. 50). The samples cover the brief interval of 12,600–12,200 yr B.P. There are no significant faunal changes in the sequence. Rather, the faunas indicate climatic amelioration from the oldest assemblage onward, with summer temperatures estimated at 14–15°C. Lemdahl contrasted this early warming with the fossil records from southern Sweden, in which assemblages of this age show summer temperatures of only 10–12°C. In fact, late glacial warming in Scandinavia began about 500 years after warming in Poland. Deglaciation took place much earlier in Poland (circa 18,000 yr B.P.) than in Scandinavia. Late-lying ice apparently kept southern Scandinavia colder than ice-free regions at similar latitudes. The timing of amelioration recorded from Zabinko is essentially synchronous with that recorded from Britain, France, and Switzerland. However, the British warming was more dramatic than that elsewhere in Europe, probably because of the overwhelming effects of sea surface temperature increases on these islands.

Nazarov (1984 and written communication, 1991) has done extensive work on insect fossil assemblages dating back to the late Cretaceous in Byelorussia and adjacent Russia. I review here only his Quaternary studies. I provide more details in this section than for other European studies because most of this literature is in Russian and not easily obtained outside Russia. The Quaternary assemblages are from sites at Artuki, Logoza, the Nizhninsky Rov Ravine, Rubezhnitsa, and Timoshkovichi (Fig. 8.5, Nos. 2, 28, 35, 41, and 45).

Cold intervals in the early Pleistocene brought the first boreal species to Byelorussia. Beginning about 730,000 years ago, at least five Pleistocene glaciations occurred; during three stadials the region was almost completely ice covered. Regional landscapes during these glaciations were dominated by periglacial environments with low summer and winter temperatures and reduced precipitation. During Pleistocene glaciations, fossil insect data suggest that summer temperatures were about 10–13°C and winter temperatures were as low as –32°C. The beetle fauna associated with periglacial environments includes species associated with tundra, forest-tundra, and montane steppe habitats. The tundra-associated species include *Diacheila arctica, Pterostichus tundrae,* and *Helophorus obscurellus.*

During the early and middle Pleistocene interglacials, warm climates allowed the return of forests and thermophilous beetles. Glacial stadial environments were extremely cold and continental, and cold-adapted Asiatic beetle species were found in these deposits. Middle Pleistocene interglacial faunas are among the richest of the Quaternary, composed of species still living today in Byelorussia.

The long Berezina Glaciation began about 440,000 years ago, as evidenced by the return of tundra species, including *Diacheila polita, D. arctica,* and

Bembidion dauricum. Even after glacial retreat, periglacial conditions persisted for a long time, and during this interval the tundra ground beetles in the *Pterostichus (Cryobius)* group first appeared in Byelorussian deposits.

The Dnieper Glacier covered almost all of Byelorussia, but the glacial advance was preceded by a lengthy cold period in which the arctic rove beetle *Tachinus arcticus* first appeared in the fossil record of this region. This species is an important element in cold stage faunas from Beringia. The faunal assemblages from the Dnieper Glaciation are a mixture of arctic and steppe species; the latter live today in Central Asia.

The Eemian Interglacial, known in Byelorussia as the Murava Interglacial (see Table 9.2 for late Quaternary correlations and chronology), fostered the most thermophilous beetle fauna of the Quaternary. This fauna was enriched by southern European species. In addition, some species from the Soviet Far East were able to penetrate Central Europe.

The first cold stage of the last glaciation began about 115,000 years ago, when such beetles as *Elaphrus splendidus, Pterostichus pinguedineus, Pterostichus tundrae, Bembidion dauricum, Patrobus septentrionis,* and *Amara torrida* arrived from northern Europe and Siberia, replacing thermophilous species. The timing of this faunal change in Byelorussia was synchronous with climatic cooling signaled by insect faunas in Britain and France. This stage was followed by the Early Poozerje Interstadial Complex. Thermophilous beetles returned to Byelorussia during the warmest part of the interstadial, in which climates were similar to those today.

During the stadials of the last glaciation several cold-tolerant species inhabited Byelorussia, including *Tachinus arcticus, Chrysolina septentrionalis, Chrysomela tajmyrensis,* and *Trichalophus korotyaevi.* The Mid-Poozerje Interstadial Complex, which was colder than the previous interstadial, allowed the penetration of new tundra species into the region.

The interval 21,000–17,000 yr B.P. was characterized by a beetle fauna enriched by Central Asian steppe species such as *Stephanocleonus eruditus, Pterostichus (Derus) majus,* and some taiga zone species. The last (late Weichselian) cold stage was more prolonged; the Poozerje Glacier advanced to cover the whole of northern Byelorussia by about 17,000 years ago. Shrub-associated and arctic tundra beetle species dominated the fossil assemblages from the beginning of this glacial interval onward. The tundra ground beetles, *Pterostichus (Cryobius), Pterostichus vermiculosus, P. sublaevis, P. tundrae, P. haematopus,* and *Amara alpina* characterize the fauna of the last glacial stadial.

During the late glacial, the Alleröd interstadial saw the brief return of a boreal fauna to Byelorussia. During the Younger Dryas interval, some tundra species returned, including *Amara alpina* and *Patrobus septentrionis.*

In the Holocene, the open ground fauna was replaced by coniferous forest species. Thermophilous species became established by 8000 yr B.P., indicating warm, moist climates. At 5500 yr B.P., the climate became warm and dry, and steppe species appear in the fossil assemblages. Cooler climate in the late Holocene brought the return of boreal species.

Bidashko and Proskurin (1988) studied middle Pleistocene insect fossils from the Nikol'skoye site on the Volga River in western Russia (Fig. 8.5, No. 34). The faunas represent a forest-steppe environment, similar to that found today in forest-steppe regions of western and central Siberia. They indicate that the study region was a mosaic of meadows, meadow-steppe, and steppe habitats, interspersed with coniferous forests.

Because British research has produced such a large body of Quaternary insect fossil data, results from continental Europe are most frequently compared with those from Britain. These comparisons are perhaps not the most fair, because the climate in Britain is and has been greatly affected by oceanic currents, whereas the interior regions of Europe have experienced continental climates. Nevertheless, the degree of similarity between British and continental insect records is gratifying, even if some of the details are different. The important similarities between these reconstructions lie not in the exact timing of changes but in the fact that they took place in the same fashion. That is, the fossil insect records from throughout Europe show that the various transitions between major climatic episodes were very rapid indeed. As far as ecosystems are concerned, the changes could be said to be catastrophic. These results support the results of the British research completely.

More extensive Quaternary insect studies have been carried out in Siberia, and they are the topic of the next chapter.

9

SIBERIAN STUDIES

> The differences in thermal regime between north- and south-facing slopes are equivalent to differences in climate on a latitudinal scale between the steppes of Central Asia and the arctic tundra of Siberia.
> —A. V. Alfimov and D. I. Berman (written communication, 1991)

During the last twenty years, Russian scientists have devoted a great deal of effort to the study of Quaternary environments of Siberia. This vast, remote region, the size of a continent, has grudgingly yielded its secrets to a few dedicated scientists. Most fossil studies have been carried out along the banks of major rivers (Figs. 9.1 and 9.2). Riverbank exposures have attracted researchers for two reasons: (1) they can be reached by boat during the summer months and (2) the banks cut by the rivers offer access to extensive exposures of Pleistocene organic-rich silts and sands. Some of these exposures contain unconsolidated sedimentary sequences that span millions of years. The sediments are in permafrost, and the fossils they contain (from pollen grains to whole mammoth carcasses) are extremely well preserved. Since the early Holocene, these exposures have thawed and eroded little by little through successive summers, gradually yielding their frozen treasure trove of fossils.

Most writings about Siberian paleoentomology have appeared in Russian publications not easily obtained outside that country. A few review articles and book chapters have been published in English. Some Russian colleagues have also provided me with translated summaries of their Russian papers. This chapter is based primarily on these summaries and English-language publications (Kiselyov, 1973; Kiselyov and Nazarov, 1984). As with the eastern European sites discussed in Chapter 8, I present more details (when available) here than I have for western European or American studies, which are more readily obtained by the interested reader. The study sites are listed in Table 9.1, and the geographic positions of the localities are shown in Figures 9.1 and 9.2.

The Siberian research is fascinating, revealing major exchanges of insect species between the western sector of Beringia and central Asia, Europe,

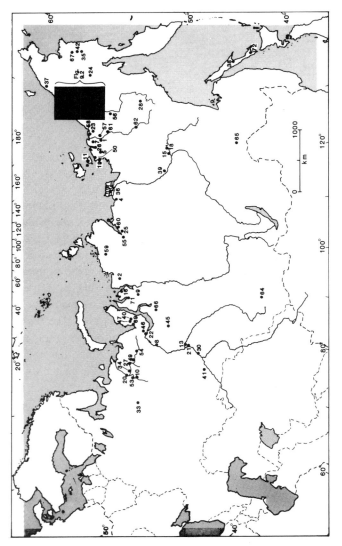

Figure 9.1. Map of Russia showing location of insect fossil localities in Siberia. The inset box is shown in the larger-scale map in Fig. 9.2. Site numbers are keyed to the list of sites in Table 9.1.

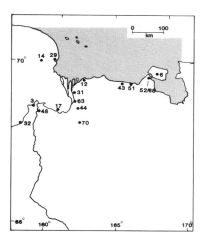

Figure 9.2. Map of northeastern Siberia showing location of insect fossil sites. Site numbers
are keyed to the list of sites in Table 9.1.

and Eastern Beringia. Siberia appears to have been the principal region in
which steppe-tundra developed and in which small patches still occur.

WESTERN SIBERIAN RESEARCH

The southernmost site of which a study has been published in English
(Kiselyov, 1973) is on the banks of the Tura River (Fig. 9.1, No. 64). A
small insect fauna, dating from the middle of the last glaciation (circa
40,000–60,000 yr B.P.) was sampled from a riverbank exposure. The assemblage was a mixture of steppe, tundra or forest-tundra, and boreal forest
(taiga) species. The tundra/forest-tundra element includes three ground
beetle species: *Nebria nivalis, Pelophila borealis,* and *Diacheila polita.*
Approximately one-third of the identified species are indicative of dry
steppe environments. There is no modern analogue for either the beetle
assemblage or the environmental conditions it reflects. Kiselyov concluded
that the fauna represents a time of rapid climatic change, in which elements
of various biological communities shifted distributions across the landscape. I discuss the phenomenon of steppe and tundra insect communities
further in the next chapter, because this combination appears repeatedly in
Siberian (and Alaskan) Pleistocene assemblages.

On the Jamal Peninsula in northwestern Siberia, Erochin and Zinovjev
(1991) have studied late glacial insect faunas from the Ljabtosjo and Ngojun
sites (Fig. 9.1, Nos. 40 and 46), and Kiselyov (1988) examined insects from
Ust-Yuribey (Fig. 9.1, No. 69). The late glacial assemblages reflect arctic

Table 9.1

Siberian Quaternary insect fossil sites and references

Site	Reference(s)
1. Achchagyy-Allaikha	Kiselyov and Nazarov (1984)
2. Agapa	Kiselyov and Nazarov (1984), Kiselyov (1988)
3. Aleshkina Zaimka	Kiselyov and Nazarov (1984)
4. Anabar River	Golosova et al. (1985)
5. Ary-Mas River	Golosova et al. (1985)
6. Ayon Island	Kiselyov and Nazarov (1984), Golosova et al. (1985)
7. Berelekh	Kiselyov and Nazarov (1984)
8. Berezovo	Kiselyov (1988)
9. Bol'shiye Kheta	Kiselyov (1988)
10. Bol'shoy Aranets	Kiselyov and Nazarov (1984)
11. Bol'shoy Lyjakhovsky Island	Golosova et al. (1985)
12. Cape Letyatkin	Golosova et al. (1985)
13. Chembakcheno	Kiselyov (1988)
14. Chukoch'ya River	Matthews (1974a), Kiselyov and Nazarov (1984)
15. Chukskoye Exposure	Kiselyov and Nazarov (1984)
16. Dorofeyevskaya	Kiselyov (1988)
17. Duvanny Yar	Giterman et al. (1982)
18. Dygdal	Kiselyov and Nazarov (1984)
19. Entrykaysky Ravine	Kiselyov and Nazarov (1984)
20. Garevo	Kiselyov and Nazarov (1984)
21. Gornopravdensk	Kiselyov (1988)
22. Gorny Kazymsk	Kiselyov (1988)
23. Keremsit River	Kiselyov and Nazarov (1984), Golosova et al. (1985)
24. Kergoli River	Golosova et al. (1985)
25. Khatanga	Kiselyov and Nazarov (1984)
26. Khroma River	Kiselyov and Nazarov (1984), Krivolutsky and Druk (1990)
27. Kipiyevo	Kiselyov and Nazarov (1984)
28. Kirghilyakh River	Kiselyov and Nazarov (1984), Krivolutsky and Druk (1986)
29. Kon'kovaya	Kiselyov and Nazarov (1984), Golosova et al. (1985)
30. Koshelevo	Kiselyov (1988)
31. Kray Lesa	Kiselyov and Nazarov (1984)
32. Krestovka River	Kiselyov and Nazarov (1984), Sher et al. (1977)
33. Kur'yador	Kiselyov and Nazarov (1984)
34. Kushshor	Kiselyov and Nazarov (1984)
35. Ledovyy Obryv	Kiselyov and Nazarov (1984)
36. Lena River	Golosova et al. (1985)
37. Leningrad	Kiselyov and Nazarov (1984)
38. Leonidovka	Kiselyov and Nazarov (1984)
39. Lepiske	Kiselyov and Nazarov (1984)

Table 9.1 (*Continued*)

Site	Reference(s)
40. Ljabtosjo	Erochin and Zinovjev (1991)
41. Malkovo	Kiselyov and Nazarov (1984)
42. Mamontov Obryv	Kiselyov and Nazarov (1984)
43. Melkera River	Kiselyov and Nazarov (1984), Golosova et al. (1985)
44. Molotkovsky Kamen'	Kiselyov and Nazarov (1984)
45. Nadym	Kiselyov (1988)
46. Ngojun	Erochin and Zinovjev (1991)
47. Nyamu	Kiselyov (1988)
48. Omolon River	Kiselyov and Nazarov (1984), Golosova et al. (1985)
49. Os'van'	Kiselyov and Nazarov (1984)
50. Oyogossiky Ravine	Kiselyov and Nazarov (1984)
51. Primorsky	Kiselyov and Nazarov (1984)
52. Rauchua River	Golosova et al. (1985)
53. Rodionovo	Kiselyov and Nazarov (1984)
54. Rogovaya	Kiselyov and Nazarov (1984)
55. Romanikha	Kiselyov and Nazarov (1984)
56. Sededema	Kiselyov and Nazarov (1984)
57. Shamanovo	Kiselyov and Nazarov (1984)
58. Shandrin River	Grunin (1973), Kiselyov and Nazarov (1984)
59. Shrenk	Kiselyov and Nazarov (1984)
60. Syndassko Bay	Golosova et al. (1985)
61. Synopy Ravine	Kiselyov and Nazarov (1984)
62. Tirekhtyakh	Kiselyov and Nazarov (1984)
63. Trety Ruchey	Kiselyov and Nazarov (1984)
64. Tura River	Kiselyov (1973)
65. Tynda	Kiselyov and Nazarov (1984)
66. Urgenoi	Kiselyov (1988)
67. Ust'-Algansky	Kiselyov and Nazarov (1984)
68. Ust'-Rauchua	Kiselyov and Nazarov (1984)
69. Ust-Yuribey	Kiselyov (1988)
70. Utlinsky Kamen'	Kiselyov and Nazarov (1984)
71. Yuribey River	Kiselyov and Nazarov (1984), Krivolutsky and Druk (1986)
72. Yuzhno-Sakhalinsk	Golosova et al. (1985), Krivolutsky and Druk (1986)

tundra conditions, similar to those found today on the Taimyr Peninsula. The assemblages included *Carabus truncaticollis, Amara glacialis, Pterostichus tundrae,* and *Tachinus arcticus.*

Kiselyov (1988) and Kiselyov and Nazarov (1984) also studied a series of Pleistocene assemblages from exposures along the Ob River (Fig. 9.1, Nos. 8, 13, 21, 22, 30, 41, and 46) and the Gyda (Fig. 9.1, Nos. 9, 16, and 71) and Taimyr peninsulas (Fig. 9.1, Nos. 2, 55, and 59) in western Siberia. During interglacial and interstadial intervals of the middle to late Pleistocene, insect assemblages suggest summer temperatures averaging 2–3°C warmer than those at present. Summer temperatures during glacial stadials were about 10°C cooler than at present, and an arctic tundra biota dominated the region. Both mesic tundra (e.g., *Pterostichus* [*Cryobius*] species) and xeric tundra beetles have been identified from stadial assemblages, in addition to some weevils found in steppe habitats.

EASTERN SIBERIA

The most thoroughly studied region is northeastern Siberia, especially the Kolyma lowland (Fig. 9.2). Regional Quaternary faunas are composed of tundra, forest-tundra, and steppe species. The steppe beetles are mostly weevils in the subfamily Cleoninae. The steppe species in the fossil assemblages live today in regions farther south in Siberia and in Mongolia. However, some species still persist in arid habitats in northern Siberia, from the Taimyr Peninsula eastward. Most of northeastern Siberia was unglaciated during the Pleistocene, but this region experienced cold, dry, continental climatic conditions during glacial stadials. Regional insect assemblages from these intervals were dominated by steppe species, in combination with beetles associated with xeric tundra habitats, including the ground beetle species *Amara alpina* and *A. glacialis,* the leaf beetle species *Chrysolina septentrionalis* and *C. subsulcata,* and the pill beetle species (Byrrhidae) *Morychus viridus* (Kiselyov and Nazarov, 1984).

Eastern Siberia formed the western part of the larger unglaciated arctic landmass called Beringia (Fig. 9.3). Beringia was a refuge for cold-adapted biota during the Pleistocene, when most of the Arctic was covered with ice. Beringia comprised unglaciated regions of eastern Siberia, Alaska, the western part of the Yukon Territory, and the exposed continental shelf between the continents.

Modern Comparative Studies
Berman (1990) studied Pleistocene records of the pill beetle *Morychus* from Siberian deposits. Pill beetle (Byrrhidae) fossils play an important part in many

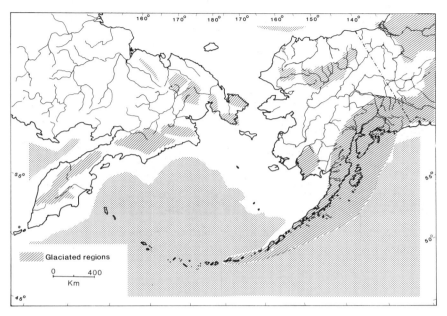

Figure 9.3. Map of the Beringian environment at about 18,000 yr B.P., showing exposed regions of continental shelf and position of glacial ice. (After Barry, 1982.)

Quaternary assemblages throughout Beringia, comprising as much as 60% of some fossil assemblages. Because of their importance to paleoecological studies, Berman conducted modern ecological research on pill beetles in eastern Siberia. Although almost all pill beetles are moss-feeders (although not necessarily in wet habitats), Berman found that one of the important species in Quaternary assemblages, *Morychus viridis,* does not live today in mesic and wet or semiaquatic moss habitats (such as bogs, swamped woodlands, and shoreline localities). Rather, it is found in xeric plant communities, such as relict steppes, mountain tundra (mostly on south-facing slopes), and various lichen-dominated communities in tundra-steppe, in places where the sedge *Carex argunensis* is present. This sedge species grows in well-drained sites, which are confined regionally to the axial parts of mountains and foothills, steep slopes, plateaus, and other similar localities (Fig. 9.4). It is a member of various plant community types. The larvae of *M. viridis* are generally confined to dense patches of the moss, *Polytrichum piliferum,* on which they live and feed.

The habitat preferences of *Morychus aeneus* differ sharply from those of *M. viridis*. *M. aeneus* is found on Kolyma River terraces in meadow localities with sandy loam soils and abundant *Ceratodon purpureus* moss.

Figure 9.4. *Morychus viridis* habitat on a dry, south-facing slope exhibiting steppe vegetation, Kolyma lowland. (Photograph courtesy Daniel Berman.)

Observations on the ecology of *M. viridis* lead to the following environmental correlations:

1. Scarcity of snow and strong winds in winter.
2. Dry summer conditions, ranging in temperature from cold to locally warm (microclimatic steppe).
3. Dominance of a mosaic vegetation cover of various groups, including sedges and mosses.
4. Variations in the ratio of steppe to tundra plants, depending on the availability of heat and permafrost conditions in microhabitats.

How do steppe-adapted species survive today in northeastern Siberia? A. V. Alfimov and D. I. Berman (written communication, 1991) have studied modern communities of relict steppe in this region. Isolated biotic communities associated with warm, dry environments occupy microhabitats on south-facing slopes on the arctic tundra. The principal ranges of these taxa are far to the south of the tundra biome in Asia and southeastern Europe, and their presence in Siberia makes for an interesting problem in modern as well as paleobiology because the fossil record demonstrates their presence in northeastern Siberia throughout much of the Pleistocene.

Two main hypotheses have been proposed for the existence of relict steppe in northern Siberia. Species of the steppe biota either persist in the north because they are adapted to cold climates or are confined to specific microclimates formed within steppe habitats that differ sharply from those of the adjacent tundra regions. The second hypothesis postulates either that the thermal conditions in the relict steppe communities are similar to those found in the southern steppes of Central Asia or that thermal conditions are not an important factor determining the distribution of these beetles.

Alfimov and Berman studied the biota, microclimate, and soil conditions on both north- and south-facing slopes in the Kolyma lowland. The north-facing slopes are consistently colder and the active layer (the layer that thaws seasonally) is considerably thinner than on the south-facing slopes, where steppe vegetation persists in patches. The soils at the steppe site thaw deeply in summer, and are well enough drained to create xeric habitats. The soils on the north-facing slope remain moist and support mesic- to moist-tundra vegetation. The study showed that, by June or early July, soil temperatures at 20 cm depth on the south-facing slope average 25–26°C and soil surface temperatures reach 58–62°C. On the north-facing slope, soil temperatures at 20 cm depth reach only 1–2°C, and the water content of the soil approaches saturation. By mid-September the soils on the north-facing slope are frozen, whereas the soils on the south-facing slope remain thawed as late as mid-October. The steppe microclimate is very hot and dry in comparison with surrounding tundra environments, and the relict steppe communities survive in the north because of the stability of the microclimate established on south-facing slopes.

Alfimov and Berman suggested that the faunal similarities between early Pleistocene thermophilous insect assemblages and modern steppe faunas indicate that throughout at least the last 2.5 million years some steppe habitats have persisted in northeastern Siberia, and that the combination of tundra and steppe species in Pleistocene assemblages may be the result of taphonomic integration, i.e., the formation of deposits that mix the remains of insects and plants from both types of habitats.

Pleistocene Insects of the Kolyma Lowland

Conditions suggestive of steppe-tundra began to take shape in the Kolyma lowlands before 700,000 yr B.P. (Giterman et al., 1982). Small patches of dry tundra expanded to large steppelike landscapes by the last glaciation, probably under the influence of very cold, dry, continental climates. This continentality was fostered by the closing off of oceanic circulation between the Pacific and Arctic oceans when sea level fell, exposing the broad continental shelf regions between the continents (Barry, 1982). This created the Bering Land Bridge, which was in existence for much of the late Pleistocene (Matthews, 1982).

During interglacial and interstadial warm periods, the steppe-tundra assemblages were replaced by boreal forest and mesic tundra assemblages, the

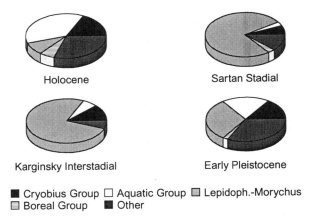

Figure 9.5. Percentage composition of ecological groups in fossil insect assemblages from Holocene (Duvanny Yar Sample 13-16), Sartan Stadial (Duvanny Yar Sample 7-8), Karginsky Interstadial (Duvanny Yar Sample 3), and early Pleistocene (Krestovka River Sample 67) deposits in eastern Siberia. (Data from Giterman et al., 1982.)

latter including species in the *Cryobius* group of the ground beetle genus *Pterostichus*. Sher et al. (1977) studied early to mid-Pleistocene insect faunas from an exposure on the Krestovka River in the Kolyma lowland (Fig. 9.2, No. 32). No radiocarbon dates are available from the Krestovka section, and much of it is undoubtedly well beyond the range of ^{14}C dating (Giterman et al., 1982). Beetles identified from early Pleistocene assemblages include both mesic and xeric tundra species, but with large numbers of *Morychus aeneus,* in addition to some steppe weevils. Figure 9.5 shows the percentage of ecological groups in this early Pleistocene assemblage, based on categories established by Matthews (1983) for eastern Beringian faunas. The compositions of Pleistocene beetle faunas from eastern and western Beringia differ significantly. Western Beringian faunas reflecting cold, dry conditions are dominated by xeric-adapted byrrhids (e.g., *Morychus*) and steppe weevils now found in southern and central Asia, whereas eastern Beringian faunas indicative of similar conditions include large numbers of the weevil species *Lepidophorus lineaticollis* and rove beetles in the *Tachinus* (*apterus*) group. Whereas *Tachinus* fossils are found in Siberian assemblages, *Lepidophorus* fossils thus far have not been found there, and in modern Siberia they are known only from the Chukotka region (Giterman et al., 1982). Conversely, Asiatic steppe weevils apparently did not become established in eastern Beringia. The *Lepidophorus-Morychus* group reflects xeric habitats.

The *Cryobius* group reflects mesic tundra environments. This group is moderately well represented in the early Pleistocene fauna from Krestovka. In eastern Beringia, *L. lineaticollis* has been found in large numbers in some

Pleistocene deposits from interior localities. The *Lepidophorus-Morychus* group is well represented in the early Pleistocene from Krestovka, a finding that shows that, even as much as 700,000 yr B.P., steppe environments were beginning to form in western Siberia (Giterman et al., 1982). However, the relative abundance of the *Cryobius* and aquatic groups shows that regional landscapes contained a mixture of xeric, mesic, and wet habitat types.

The upper section of the early Pleistocene reflects more mesic conditions. The beetle faunas include water beetles, boreal carabids, few *Morychus,* and no steppe weevils.

Mid-Pleistocene environments have been described as the Olyor Suite in northeastern Siberia. An exposure on the Chukoch'ya River (Fig. 9.2, No. 14) yielded peats studied by Matthews (1974a). He identified a small fauna (30 taxa), including *Pterostichus* (*Cryobius*) species, *Amara alpina, Trichocellus punctatellus,* and abundant *Morychus* specimens. Again, these elements combine in a steppe-tundra environment. The eastern Beringian fauna with the greatest similarity to the Olyor Suite is the late Pleistocene fauna from Cape Deceit, Alaska (Chapter 10). This fauna also suggests steppe tundra, but roughly contemporaneous (mid-Pleistocene) assemblages from Cape Deceit indicate mesic tundra, not steppe-tundra.

At the Krestovka River site, Olyor Suite faunas and floras once again indicate steppe-tundra environments. The beetle assemblages include steppe weevils, *Tachinus apterus,* and *Bembidion dauricum,* but also carpenter ants of the species *Camponotus herculeanus,* which needs trees to make nests. Younger Olyor Suite assemblages from Krestovka show increased numbers of boreal forest insects, including the boreal ground beetle species *Trachypachus zetterstedti* as well as *C. herculeanus.* These results appear to differ from those for Matthews's Olyor Suite faunas; the species composition is of a very different character. It is likely that the sites are not precisely contemporaneous. After all, the term "mid-Pleistocene" covers many tens of thousands of years, and, except for extinct mammalian faunas, few chronostratigraphic markers establish the age of these Siberian deposits.

The late Pleistocene (Sartanian) is represented in northeastern Siberian fossil assemblages from the Omolon River (Fig. 9.2, No. 48; Fig. 9.6). Preliminary analyses are discussed in Kiselyov and Nazarov (1984). I have examined additional samples, made available by Dr. André Sher, dating to the Karginsky Interstadial (see Table 9.2 for correlations) in the middle of the last glaciation (circa 31,000 yr B.P.). The beetles include a mixture of xeric tundra taxa, such as *Amara alpina, Cymindis, Trichocellus punctatellus,* and abundant *Morychus,* in conjunction with mesic tundra forms, such as several species of *Pterostichus* (*Cryobius*). The overall impression given by the fauna is of fairly

Figure 9.6. A Quaternary exposure on the Omolon River, showing the study site. Note the massive ice wedges. (Photograph courtesy Mary Edwards.)

xeric tundra with some mesic and wet habitats available. The mid-Sartanian fauna is most like faunas identified from interior sites in Alaska and the Yukon (see Chapter 10).

Giterman et al. (1982) discussed Karginsky Interstadial faunas from the Duvannyy Yar site (Fig. 9.2, No. 17). The ecological composition of this fauna is shown in Fig. 9.5. Mesic tundra and aquatic elements are overshadowed by xeric or steppe species. There is no evidence of regional tree establishment during this interstadial.

During the height of the Sartan Stadial, the Duvannyy Yar faunas were dominated more completely by xeric-adapted beetles; mesic tundra and aquatic species are at a minimum for the Pleistocene (Fig. 9.5). A Holocene assemblage from this site shows a recovery of mesic tundra and aquatic species, at the expense of xeric taxa. It also contains some boreal species.

Golosova et al. (1985), Krivolutsky and Druk (1986), and Krivolutsky et al. (1990) discussed the Quaternary oribatid mite fauna of Siberia. A new method of extraction of fossil oribatid mites from Quaternary paleosols and peats is presented in these papers. The oribatids are classified in five adaptive types: inhabitants of the soil surface, of deep soil, and of small soil holes; nonspecialized forms; and aquatic oribatids. Siberian studies include investigations of

Table 9.2

Correlation of Eastern European and Siberian Pleistocene sequences with oxygen isotope stages

Isotope stage[a]	Approximate age (yr B.P.)[b]	Byelorussian sequence[c]	European Russian sequence[d]	Siberian sequence[e]
1	0–10K	Holocene	Holocene	Holocene
2	10–22K	Late Poozerje Glaciation	Late Valdai Glaciation	Upper Zyrianka/ Sartanian Stadial
3	22–55K	Mid-Poozerje Interstadial Complex	Bryansk/Dunaevo Interstadial	Karginsky Interstadial
4 } 5a }	55–110K	Early Poozerje Stadial (began circa 115K yr B.P.)	Early Valdai Stadial	Lower Zyranka Stadial
5d				
5e	110–130K	Murava Interglacial	Mikulino Interglacial	Kazantsevo Interglacial
6	130–180K		Dnieper Glaciation	Taz Glaciation

[a] Isotope stages after Shackleton and Opdyke (1973).
[b] Ages after Arkhipov et al. (1986).
[c] Byelorussian correlations after Nazarov (written communication, 1991).
[d] Eastern European correlations after Velichko and Faustova (1986).
[e] Siberian correlations after Arkhipov et al. (1986).

fossil faunas from peat samples on Sakhalin Island (Fig. 9.1, No. 72) and northern Siberia, including the mammoth carcass localities on the Kirghilyakh and Yuribey Rivers (Fig. 9.1, Nos. 28 and 71). Only one oribatid species was found in the fossil assemblages that is not a part of the recent Siberian fauna. The mite fauna of the Olyor Suite (mid-Pleistocene) was indicative of rather severe climates and tundra environments similar to those of modern eastern Siberia. Holocene mite faunas were the most diverse and abundant of the Quaternary.

Grunin (1973) described an extinct species of bot fly associated with a mammoth carcass found on the Shandrin River in northeastern Siberia (Fig. 9.1, No. 58). Larval exuviae of *Cobboldia russanovi* were recovered from the mammoth carcass. Some species of bot flies are known to parasitize modern elephants, but these fossils do not match any known extant species, so perhaps the species they represent became extinct along with their host.

Regional Paleoenvironmental Syntheses

As we have seen, Russian research in Siberia offers a unique synthesis of fossil and modern biological studies. S. V. Kiselyov (written communication, 1991) has synthesized the paleoenvironmental interpretations drawn from the Siberian insect fossil studies. In Siberia, as elsewhere, ecosystem structure depends on the dynamics of environmental change. The fossil record reveals a high level of structural stability in the biotic communities of northeastern Siberia. During the late Pleistocene, latitudinal differences in response to environmental changes over this vast region were slight. The dominant climatic factor affecting biotic communities in the north was probably precipitation rather than temperature.

The vast Siberian lowlands, including the Kolyma lowland, were unglaciated during the Pleistocene, and the land area exposed by lowered sea level increased. In the east and partly also in the south, vast coastal plains were surrounded by mountains, which completely restricted Pacific maritime influences on climate.

Based on the steppe elements in the faunas, Kiselyov estimated that mean July temperatures in the late Pleistocene of northeastern Siberia were 14–15°C. Climate changed only slightly until postglacial times. The climate approached modern parameters by 11,000–10,000 yr B.P. On the modern arctic coast of western Chukotka, mean summer temperatures reached 12°C (they are now 6°C). This warming event has also been registered by contemporaneous insect fossil assemblages from the exposed Chukchi Sea shelf (Elias et al., 1992a) and the Arctic Coastal Plain of Alaska (Nelson and Carter, 1987). Insect fossils from late glacial assemblages in western Siberia indicate that the climate was essentially modern there by 11,000 yr B.P.; there is no evidence of a Younger Dryas cooling in these insect fossil records.

In the early Holocene, paleoenvironmental reconstructions based on insect data show that climate was the main factor affecting Siberian biotic communities. Evidence for latitudinal differences in climatic regime and biotic provinces is slight, in marked contrast to today. Kiselyov hypothesized maximum Holocene warming from 9000 to 8000 yr B.P., based on range expansions of such thermophilous beetle taxa as the chrysomelids *Donacia* and *Plateumaris*.

On the northern coastal plains, an essentially modern climatic regime was established by 7000 yr B.P.

Evidence from the Jamal region (Erochin and Zinovjev, 1991) suggests that spruce-larch forests spread across valleys by 6000 yr B.P., in response to a climatic amelioration of 1–2°C. The composition of regional beetle faunas also suggests July temperatures at modern levels at this time.

It appears that the climatic regime of northeastern Siberia was similar to that of modern times by 7000 yr B.P., but in regions west of the Lena River temperatures were warmer than modern ones by 1–2°C. This finding also holds true for the region west of the Ural Mountains.

The Russians and Byelorussians have mounted serious research efforts in Quaternary entomology during the past twenty years, concentrated in a few study regions. Although only a small portion of their work has been published in English, the available literature offers tantalizing bits and pieces of a truly enormous puzzle. One continuing problem is that their work stands in almost complete isolation from correlative research elsewhere in Europe and across the Bering Strait in Alaska. For instance, it seems very likely that many species described from eastern Siberia are synonymous with Alaskan and northern Canadian species, but this question cannot be resolved until taxonomists have access to specimens from all of these regions. It is to be hoped that the linguistic and geographic isolation that Rusian scientists suffered during the last few decades will now diminish and that we will be able to share ideas, specimens, and field trips more readily in the future.

The next chapter reviews insect fossils and Quaternary environments from across the Bering Straits in eastern Beringia.

10

EASTERN BERINGIAN STUDIES

> How far into the past must we probe to understand the natural periodicities of
> world climate? I believe we must look back at least to the late Tertiary.
> —John Matthews (1989b)

In this chapter, I pick up the threads of the Beringian story as they have
developed on the eastern side of the arctic refugium. The origins of the modern
arctic ecosystems date back to late Tertiary environments in this region. To put
the Beringian story into its proper context, I begin here with a summary of
regional environmental history during that period.

LATE TERTIARY ORIGINS OF THE TUNDRA BIOME

The frozen sediments of the high arctic are the repository for remarkable late
Tertiary fossils, which are exceptional for three reasons:

1. The permafrost has preserved them exceedingly well for several
   million years.
2. The fossils represent either extant species or their immediate
   precursors (see Chapter 4).
3. They offer us unusual insights into the origins of northern
   ecosystems.

Matthews (1979a,b, 1981, 1989a,b) has written excellent reviews of these late
Tertiary faunas. The fossil beetle and plant macrofossil evidence indicates that
the arctic tundra biome originated sometime in the late Tertiary, before the
Pleistocene glaciations (Matthews, 1979a,b). Other late Tertiary deposits contain boreal forest insects, including sites in the high arctic. Late Tertiary insect
faunas have also been described from regions outside Beringia, including the
Canadian arctic archipelago and northern Greenland. Table 10.1 provides a list
of key arctic late Tertiary insect fossil sites.

Table 10.1

Arctic late Tertiary insect fossil sites and references

Site	Reference(s)
1. Banks Island (multiple sites)	Matthews et al. (1986)
2. Borden Island	Matthews (1989b)
3. Colville River	Morgan et al. (1979)
4. Ellef Ringnes Island	Matthews (1989b)
5. Haughton Astrobleme	Matthews (1989b)
6. Kap København	Böcher (1989a, 1990)
7. Lava Camp Mine	Matthews (1970), Hopkins et al. (1971)
8. Meighen Island	Matthews (1974d, 1976b, 1977a, 1989a)
9. Prince Patrick Island	Matthews et al. (1990c)
10. Strathcona Fiord	Matthews (1989b)

The Lava Camp Mine is a western Alaskan late Tertiary site examined for insect fossils (Matthews, 1970; Hopkins et al., 1971). This site contained woody, peaty alluvium and pond sediments capped by a volcanic basalt. The basalt has been dated at 5.7 million yr B.P. The insect fauna is similar in composition to the modern fauna of the Pacific Northwest coniferous forests. This fauna included extinct forms of the rove beetle *Micropeplus* that are ancestral to the modern species.

The only other late Tertiary fauna described from Alaska was extracted from a deposit on the Colville River on the North Slope (Morgan et al., 1979). Unconsolidated organic sediments exposed in a cut bank contained fossil logs of spruce, larch, and fir. Fir fossils have been found in Miocene and Pliocene deposits in arctic North America. Insect remains were extracted from detrital woody peat. A tundra fauna was identified, including *Tachinus apterus, Pterostichus brevicornis, P. nivalis,* and *Amara bokeri.*

More extensive late Tertiary deposits, a part of the Beaufort Formation, have been found on the Canadian arctic archipelago. The best studied of the Beaufort Formation insect faunas come from Meighen Island, which today has a severe arctic climate. Some typical arctic plants and insects were found in these assemblages, including an extant species of *Diacheila* (Carabidae) closely related to *D. polita, Blethisa multipunctata,* and several species of *Cryobius.* Altogether, the fauna suggests a lowland tundra near treeline. It is perhaps the earliest insect fauna representing this environment (Matthews, 1989a).

In northernmost Greenland, Böcher (1989a, 1990) has described late Tertiary insect faunas from the Kap København Formation. The age of the formation is probably 2–2.5 million yr B.P. Frozen peats from Kap København contain a boreal flora, including large spruce twigs and cones. The insect remains are amazingly well preserved. Fossils representing boreal zone taxa include bark beetles, carpenter ants, weevils, and tiger beetles. These fossils corroborate the botanical evidence of coniferous forest at 80° north latitude.

Research continues on high arctic Tertiary deposits; this work will assuredly continue to produce many important discoveries.

EASTERN BERINGIAN QUATERNARY STUDIES

> To recreate a landscape, green with life or windswept and barren, and then repopulate it with animals and men is a formidable task.
> —Schweger et al. (1982)

Beringia has captured the imagination of many researchers, because it was essentially a world unto itself for much of the Pleistocene. A unique refuge for arctic biota, it spanned the margins of two continents and was nearly surrounded by ice. As reported in Chapter 9, western Beringia experienced cold, dry environments during much of the late Pleistocene, with steppe elements more important than mesic tundra elements in a mosaic of biological communities.

The nature of Pleistocene environments in eastern Beringia is a topic of considerable debate. Paleobotanical research has indicated that steppe-tundra was important, especially in the interior basins of Alaska and the Yukon territory (Matthews, 1982). Insect fossil studies add a new dimension to this scenario.

Macrofossil data indicate that eastern Beringia was not a uniform steppe-tundra ecosystem during the Pleistocene, but rather a mosaic of different biological communities, many of which no longer exist. For instance, recent work in southwestern Alaska indicates the persistence of mesic and moist tundra habitats, even during the height of the last glaciation.

Although about thirty publications have been based on Quaternary insect fossil studies from eastern Beringia, an enormous amount of work has yet to be done. Many regions remain untouched, and there are many temporal gaps. As in eastern Siberia, much of Alaska and the Yukon remains nearly inaccessible wilderness. Likewise, most eastern Beringian investigations have been of deposits from riverbank exposures. Many of these exposures, like their Siberian counterparts, contain organic deposits in permafrost, spanning as much as several million years. I will examine various regions of eastern Beringia in turn, starting with the Alaskan interior.

Interior Alaska

The first Alaskan Quaternary insect fossil study (Matthews, 1968) was of late Pleistocene assemblages at Eva Creek, near Fairbanks (Fig. 10.1, No. 11); it

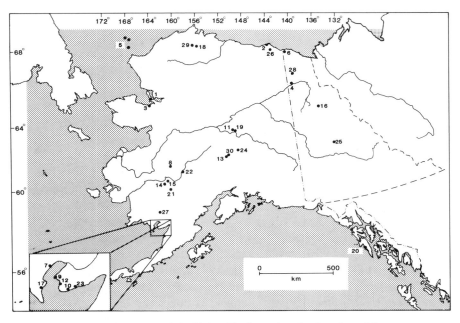

Figure 10.1. Map of Alaska and the Yukon Territory showing location of Quaternary insect fossil sites. Site numbers are keyed to the list of sites in Table 10.2.

was followed by work in the nearby Isabella Basin (Fig. 10.1, No. 19) (Matthews, 1974c). The oldest assemblage from Eva Creek is probably Sangamon Interglacial in age, based on its stratigraphic position and a radiocarbon date of >56,900 yr B.P. Another sample probably relates to a mid–Wisconsin Interstadial environment, and a third is full glacial, dated at 24,400 yr B.P. The sedimentary sequence at Eva Creek is an organic-rich silt, called "muck" by the local placer gold miners, who wash away thousands of tons of frozen muck per year to expose gold-bearing deposits underneath. This sediment is probably partly colluvial in origin, a fact that casts some doubt on whether the insects it contains are contemporaneous and on whether they represent more than one biotope.

The Eva Creek insect fauna from the Sangamon Interglacial is a mixture of mesic and xeric tundra species, with some boreal elements. The boreal zone insects are not tree-associated species (e.g., carpenter ants or bark beetles), even though there are spruce stumps in the Eva Creek Sangamon beds. The mid–Wisconsin Interstadial (?) fauna contains abundant mesic tundra species in the *Cryobius* group and some xeric tundra elements. The full-glacial fauna contains fewer *Cryobius* specimens and increased numbers of the weevil *Lepi-*

Table 10.2
Eastern Beringian Quaternary insect fossil sites and references

Site	Reference(s)
1. Baldwin Peninsula	Hopkins et al. (1976)
2. Barter Island	Wilson and Elias (1986)
3. Cape Deceit	Matthews (1974a), Giterman et al. (1982)
4. Ch'ijee's Bluff	Matthews et al. (1990a)
5. Chukchi Shelf	Elias et al. (1992a)
6. Clarence Lagoon	Matthews (1975a)
7. Coffee Point	Lea et al. (1991)
8. Colorado Creek	Elias (1992a)
9. Ekuk Bluffs	Lea et al. (1991)
10. Etolin Point	Lea et al. (1991)
11. Eva Creek	Matthews (1968)
12. Flounder Flat	Lea et al. (1991)
13. Foraker River	Waythomas et al. (1993)
14. Holukuk Mountain	Short et al. (1992)
15. Holukuk River	Short et al. (1992)
16. Hungry Creek	Hughes et al. (1981)
17. Igushik	Lea et al. (1991)
18. Ikpikpuk River	Nelson and Carter (1987)
19. Isabella Basin	Matthews (1974b)
20. Kruzof Island	Klinger et al. (1990)
21. Kulukbuk Bluffs	Elias (1992c)
22. Kuskokwim/Big River	Elias (1992c)
23. Kvichak Peninsula	Lea et al. (1991)
24. Landslide Bluff	Waythomas et al. (1989)
25. Mayo Village	Matthews et al. (1990b)
26. Niguanak Uplands	Wilson and Elias (1986)
27. Nuyakuk River	Elias and Short (1992)
28. Old Crow	Matthews (1975b)
29. Titaluk River	Nelson and Carter (1987)
30. Toklat River	Elias et al. (1993)

dophorus lineaticollis, indicative of dry habitats. Surprisingly, there was no increase in the arctic rove beetle *Tachinus apterus,* which has been found in large numbers in many eastern Beringian full-glacial faunas. Although the data from this study are interesting, the faunal assemblages of individual horizons have possibly been mixed through colluvial slope wash and reworking; the dating control of the site also remains a problem.

At Isabella Basin, a 27-m core was taken through silts and organic-rich lenses. The core sampled sediments with the same origin as those at Eva Creek, so the same caveats apply (e.g., colluvial sediments may represent mixed assemblages). The base of the core dates to the mid-Wisconsin (34,900 yr B.P.); the top of the core is mid-Holocene in age (4510 yr B.P.). Beetle assemblages were extracted from the core. A mid-Wisconsin fauna includes both mesic and xeric tundra species. Although conifer pollen was lacking from the pollen spectrum, spruce macrofossils were found. Matthews (1974b) interpreted the local environment as similar to that of modern northern treeline.

The last Wisconsin glacial fauna was depauperate in species and dominated by *Morychus* and *Lepidophorus lineaticollis*. The sediments also yielded steppe-tundra pollen spectra. Surprisingly, the Holocene fauna was also depauperate and nondescript, but xeric tundra species were not found. Modern vegetation appears to have developed by 9200–7800 yr B.P.

The Alaska Range south of Fairbanks is a broad arc of tall mountains separating the dry Alaskan interior to the north from moister regions to the south. I have recently studied some late Quaternary insect fossils from river-bank exposures on the north side of the range. These sites were unglaciated in the late Wisconsin, but the toes of mountain glaciers were close by, and the effects of these glaciers can be seen in local climatic reconstructions. I examined two sites on the Foraker River (Fig. 10.1, No. 13) (Waythomas et al., 1993). Radiocarbon dates have been obtained for three samples from one Foraker locality and six from another locality. Basal peat from the Foraker Slump site has been dated >42,000 yr B.P.; upper peat layers at the site date to the early to mid-Holocene. At the Foraker Ice Wedge locality (Fig. 10.2), organic lenses dated at >35,900 yr B.P. were frozen in silts that were deformed by massive ground ice. Organic materials filling an ice wedge cast at the site yielded late Holocene ages.

The oldest sample from the Slump site contained *Tachinus brevipennis,* a sister species to *T. apterus,* and one likewise adapted to very cold conditions. The assemblage also yielded fossil *Cryobius* specimens. The paleobotanical evidence from this horizon suggested the presence of scattered conifers, and we interpreted the landscape as an open taiga, but the cold-adapted beetles indicate the presence of open, meadow areas. Beyond the scope of radiocarbon dating, this assemblage fits stratigraphically into a mid-Wisconsin context.

The Ice Wedge locality yielded Holocene fossil insect samples. Early Holocene (8075 yr B.P.) insect data suggest a birch shrub environment, possibly with some tree birch and scattered alder, but with few conifers locally. A late Holocene organic deposit overlies a massive, frozen silt that may have been deposited during the Neoglacial period. This unit contained a boreal insect

Figure 10.2. A Quaternary exposure on the Foraker River (the Foraker Ice Wedge Site), showing massive ground ice and sediments contorted by ice wedge expansion. (Photograph by the author.)

fauna, but the continued presence of some *Cryobius* species suggests the persistence of some open ground.

Another site on the north side of the Alaska Range is the Landslide Bluff site on the Nenana River (Fig. 10.1, No. 24). Four peats were collected here; they yielded infinite radiocarbon ages but contained abundant spruce macrofossils, including logs (Waythomas et al., 1989). The peats may therefore be Sangamon in age, although spruces were apparently growing in the Fairbanks region during the mid–Wisconsin Interstadial, so that interval cannot be ruled out. The beetle fauna contained mostly northern boreal species, but also had some mesic tundra indicators. Taken together, this evidence suggests northern treeline conditions, not unlike Matthews's Eva Creek mid-Wisconsin age sample.

A third site from the Alaskan interior is the Colorado Creek mammoth site near McGrath (Fig. 10.1, No. 8). Insect fossils were extracted from organic-rich silts surrounding mammoth remains and from a mammoth dung bolus (Elias, 1992a). The fauna dates to the late Wisconsin Stadial (15,000 yr B.P.), with two components to the fossil insect assemblage: one associated with the mammoth remains and another from the surrounding upland environment. Not surprisingly, the mammoth dung contained abundant *Aphodius* dung beetles (*A. congregatus*), and the mammoth remains yielded the carrion beetle species (Silphidae) *Silpha coloradensis*. In addition, abundant blowfly pupae were found in the nasal cavities of the mammoth skull.

The small fauna representing the upland environment included the pill beetle genus *Morychus* and the leaf beetle genus *Chrysolina,* as well as a few *Cryobius* specimens in combination with *Lepidophorus lineaticollis*. This small assemblage may represent a xeric habitat developed within a broader region of mesic tundra, or it may have been part of the larger steppe-tundra biome of the Alaskan interior. If so, the geographic location of the site is critical, because all the studied sites to the south of Colorado Creek have yielded mesic and hygrophilous faunas that lack xeric-adapted taxa. The insect fossil fauna, although too small to be definitive, offers at least a suggestion that the Colorado Creek site was near the southwestern boundary of steppe-tundra environments in the late Wisconsin.

Studies in the Yukon Territory

Pleistocene exposures along the Old Crow River in the northwestern Yukon Territory (Fig. 10.1, No. 28) first won attention because of archaeological finds. In conjunction with the archaeological studies, Matthews (1975b) and Morlan and Matthews (1983) studied insect fossil remains from a sequence of late Quaternary organic deposits. The age of the insect fossils ranges from pre-Sangamon to mid-Wisconsin (32,000 yr B.P.).

The pre-Sangamon fauna was a mixture of mesic tundra and xeric open ground species with few hygrophilous taxa. Above this horizon is the Old Crow tephra, which has become an important stratigraphic marker horizon throughout much of eastern Beringia. It has been dated by various methods from about 86,000 to 149,000 yr B.P. (Wintle and Westgate, 1986; Westgate, 1988), but an overview by Hamilton (1991) placed the age at about 130,000 yr B.P.

One sample of insect fossils from the Old Crow site represents a steppe-tundra environment and came from sediments just above the tephra. Immediately above this assemblage in the Old Crow sequence is a fauna showing a decline in xeric species and a large increase in hygrophilous taxa. This fauna may represent the Sangamon Interglacial, but it lacks forest insects.

A steppe-tundra fauna dominates all Wisconsin-age assemblages, including one dated at 32,000 yr B.P. There is some indication of forest species in Old Crow assemblages by 13,500 yr B.P., followed by a drastic decline in xeric fauna and an increase in hygrophilous species in the Holocene.

Ch'ijee's Bluff on the Porcupine River is quite near the Old Crow localities (Fig. 10.1, No. 4). Matthews et al. (1990a) presented the results of fossil insect analyses from this site. The assemblages represent Sangamon and pre-Sangamon environments. Samples associated with the Old Crow tephra contain abundant remains of the arctic rove beetle, *Micralymma brevilingue,* in addition to *Tachinus apterus* and *Lepidophorus lineaticollis,* suggesting steppe-tundra conditions. This reconstruction is supported by the pollen, dominated by sedge, grasses, and birch.

The Sangamon fauna was very diverse, comprising 120 species, including bark beetles, boreal carabids, and abundant aquatic and hygrophilous species. The boreal ground beetle fauna included species living far to the south of the site today in British Columbia and the southern Yukon. This fauna is indicative of boreal forest and climatic conditions warmer than those at present. Originally, it appeared to be at odds with the fauna from Old Crow that was thought to be of Sangamon age. That fauna contained no boreal elements. However, the supposed Sangamon beds from Old Crow have now been redated (J. V. Matthews, Jr., written communication, 1992), and are now thought to be mid-Wisconsin in age. The Ch'ijee's Bluff Sangamon fauna is indicative of conditions in the heart of the boreal forest.

With the addition of forest beds from Old Crow, there are now several regions in eastern Beringia providing evidence of establishment of boreal forest during the mid-Wisconsin Interstadial, including sites in the Yukon Territory and the Alaskan interior. Sites in southwestern Alaska contain mid-Wisconsin assemblages reflecting either steppe-tundra or mesic tundra. The distribution of sites with boreal assemblages suggests a loose geographic trend, with forest establishment in interior sites and persistence of tundra elsewhere. On the other hand, published information is available only for six sites with mid-Wisconsin assemblages in the whole of eastern Beringia, so it seems likely that the data are lacking to develop a regional understanding of this period.

Matthews (in Hughes et al., 1981) also studied Quaternary insect faunas from the Hungry Creek site in the Bonnet Plume Basin (Fig. 10.1, No. 16). The age of the samples was thought to have been early to mid-Wisconsin. The insects described as an early Wisconsin fauna include only obligate or facultative tundra dwellers, such as the arctic ground beetles *Amara alpina, Carabus truncaticollis,* and *Diacheila polita.* Specimens of *Tachinus apterus, Morychus,* and *Lepidophorus lineaticollis* were also found.

An assemblage initially thought to have been mid-Wisconsin in age is now considered to represent Sangamon environments (J. V. Matthews, Jr., written communication, 1992). It contained boreal ground beetles, bark beetles, and two ground beetle species indicative of climate warmer than that at present (*Pelophila rudis* and *Chlaenius niger*). Some arctic or subarctic species were also found in this assemblage, but Matthews argued that these beetles were transported downstream from the alpine tundra to the depositional basin.

A mid-Wisconsin insect assemblage was described from a site on the Stewart River in the central Yukon, at the Mayo Indian village (Fig. 10.1, No. 25) (Matthews et al., 1990b). This steppe-tundra assemblage dates to 29,600 yr B.P. and includes four species in the *Cryobius* group, *Amara alpina, Morychus, Lepidophorus lineaticollis, Harpalus amputatus,* and *Cymindis. H. amputatus* is a grassland species, found today across the prairies of the western United States and Canada. The insect data, in combination with palynological and plant macrofossil evidence, suggest a treeless landscape, with at most small spruce groves, and summer temperatures at least 5°C colder than at present. Ice from the late Wisconsin McConnell Glaciation overrode the site shortly after this assemblage was deposited.

North Slope Alaskan Studies

Stretching north from the northern foothills of the Brooks Range to the shores of the Beaufort Sea, the North Slope of Alaska is today an arctic tundra underlain by continuous permafrost. A few peat deposits and organic lenses in silts have been studied from this region. Most of these deposits date to the Holocene or to the Wisconsin-Holocene transition. The only published mid-Wisconsin record is from the Titaluk River (Fig. 10.1, No. 29). Nelson (1986) and Nelson and Carter (1987) described assemblages ranging in age from 42,000 to 30,000 yr B.P., indicative of dry substrates and discontinuous vegetation cover. The assemblages included some typical steppe-tundra elements, including *Harpalus amputatus* and *Morychus.* The assemblages show a progression toward increasing aridity, but with summer temperatures similar to those of modern times.

The most easterly of the late Wisconsin–early Holocene sites was described by Matthews (1975a) from Clarence Lagoon, just east of the U.S.-Canadian border on the Yukon coast (Fig. 10.1, No. 6). A birch twig in the sample yielded an age of 10,900 ± 80 yr B.P. The insect assemblage represents essentially a facultative tundra fauna with some northern boreal elements, not unlike the regional fauna of today.

The Niguanak Uplands region (Fig. 10.1, No. 26) contains many thaw lakes. The banks of one of these lakes yielded peat, studied by Wilson and Elias (1986) for insect fossils. The age of the samples ranged from 10,400 to 1320 yr B.P. The insect fauna was essentially modern throughout. The assemblage from 6000 yr B.P. yielded the boreal rove beetle, *Tachyporus canadensis,* a finding that suggests climatic conditions warmer than those at present.

Peats and organic lenses in sand were sampled from a coastal bluff on Barter Island (Fig. 10.1, No. 2) (Wilson and Elias, 1986). These samples spanned the interval 7800–6700 yr B.P., but the sequence of dates was inverted, suggesting reworking. The entire interval was therefore treated as a bulk sample. The Barter Island insect fauna consisted of mesic tundra, riparian, and aquatic beetles. The paleoenvironmental reconstruction was of a thaw lake surrounded by sedge marsh, with essentially a modern climate.

Farther west on the coastal plain flows the Ikpikpuk River (Fig. 10.1, No. 18). Nelson and Carter (1987) discussed an insect fauna dating to 9400 yr B.P. The deposit also contained extralimital *Populus* macrofossils and pollen, suggestive of climatic conditions warmer than those at present. The insect fauna contained more than 100 beetle taxa and 60 identified species, including abundant tundra carabids, with ten species of *Cryobius, Diacheila polita,* and *Carabus chamissonis.* It also contained extralimital carabids, such as *H. amputatus, Blethisa multipunctata, Agonum quinquepunctatum, Agonum quadripunctatum, Amara erratica, H. ful-vilabrus,* and the byrrhid *Cytilus alternatus.* All of these are found today in the boreal zone of north-central Alaska and adjacent Canada. Although the extralimital *Populus* data might be taken to indicate only increased moisture at the site, the beetle data clearly indicate North Slope climatic conditions in the early Holocene 3–5°C warmer than at present.

Perhaps the most intriguing evidence from this region is derived not from the North Slope proper, but from terrestrial peat deposits that are now beneath 50 m of water in the Chukchi Sea. These deposits formed on the Bering Land Bridge, when the shallow continental shelf regions were exposed during the late Pleistocene. Cores taken on board research ships have yielded profiles of terrestrial peats, ranging in age from >40,000 to 11,000 yr B.P. (Elias et al., 1992a). Peats and organic-rich silts from a transect of sites on the Chukchi Shelf (Fig. 10.1, No. 5) yielded a small insect fauna, dating from 11,300 to 11,000 yr B.P. The insect fossils are indicative of arctic coastal habitats like those of the Mackenzie Delta region, suggesting that, during the late glacial, the exposed Chukchi Shelf had a climate 6–10°C warmer than the modern arctic coast of Alaska. The insect and pollen data suggest a meadowlike graminoid tundra with shrubs growing in sheltered areas.

The radiocarbon dates from the Chukchi Shelf peats provide important new data on the timing of land bridge inundation. The peats were deposited on the edge of a small pond in a coastal plain environment. Fresh water– and brackish water–adapted ostracodes were also found in the late glacial peats, suggesting that sea level—which had been well below the elevation of the site throughout much of the Wisconsin Glaciation—was rising to the level of the core sites at about 11,000 yr B.P. (Phillips and Brouwers, 1990). Estimates have placed late glacial sea level at about −70 m at 11,000 yr B.P. (Fairbanks, 1989). McManus et al. (1983) and McManus and Creager (1984) provided estimates of past sea level for the Bering-Chukchi shelves, suggesting that sea level was near −55 m by 16,000 yr B.P. However, the dates assigned to this chronology are suspect, because they are based on radiocarbon ages of bulk sediments that were probably contaminated with coal.

The radiocarbon ages of plant macrofossils from the Chukchi Shelf provide evidence that the previous dates on bulk sediments are as much as 3800 years too old. If the marine transgression took place after 11,000 yr B.P., then the timing of the transgression coincided with the marked warming seen in the North Slope insect faunas. As long as the Bering Land Bridge was in existence, it fostered increased climatic continentality in eastern Beringia. However, once the land bridge was flooded, that continental climatic regime probably broke down. Hence, the reestablishment of circulation between the Pacific and Arctic oceans through the Bering Strait probably played a major role in climatic amelioration in eastern Beringia around 11,000 yr B.P. The Chukchi Shelf evidence suggests that this effect was both rapid and intense, not unlike the amelioration seen in British insect faunas following the return of warm Gulf Stream waters after the Younger Dryas oscillation.

Studies in Western Alaska

Matthews (1974a) studied a series of Quaternary insect assemblages from Cape Deceit on the Seward Peninsula (Fig. 10.1, No. 3). Most of the assemblages are beyond the range of radiocarbon dating, and Matthews has modified his original age assessments. The horizons assigned to the Cape Deceit formation are of early Pleistocene or late Pliocene age. Insect faunas from this unit are indicative of open ground environments, possibly just beyond treeline (Giterman et al., 1982). Above the Cape Deceit Formation is the Deering Formation, the base of which was deposited during a warm interval that supported the establishment of spruce forest. This formation may represent the Sangamon Interglacial (Giterman et al., 1982). Geologic evidence suggests that this lower Deering Formation was deposited at a time when the land bridge was flooded. This flooding in turn may have contributed to the climatic amelioration that allowed the most westerly expansion of treeline in the late Pleistocene. The younger assemblages from the Deering Formation represent early Wisconsin environments and contain insect assemblages indicative of dry, open ground conditions. Insects from an assemblage dated at 12,400 yr B.P. include the typical steppe-tundra taxa *Morychus, Lepidophorus lineaticollis,* and others. The Cape Deceit assemblages are important to our understanding of the development of tundra ecosystems in eastern Beringia. The earliest assemblages document the establishment of coastal tundra, probably before the Pleistocene. The Cape Deceit data also show that Pleistocene tundra communities differed from modern tundra and that steppe-tundra biotopes developed in eastern Beringia during cold, dry episodes in the Pleistocene (Giterman et al., 1982).

Matthews (in Hopkins et al., 1976) also studied insect fossils from a mammoth site on the Baldwin Peninsula (Fig. 10.1, No. 1), just north of Cape Deceit. The mammoth and associated organic sediments were dated at about 27,000 yr B.P. Only a few insect taxa were identified, including the tundra water beetle *Helophorus splendidus,* either *Tachinus apterus* or *T. brevipennis,* oribatid mites, and a spider. The insect data, in combination with paleobotanical evidence, is indicative of shrub tundra rather than steppe-tundra.

Southwestern Alaskan Studies

Fossil assemblages from 11 late Quaternary sites in southwestern Alaska provide evidence of the survival of an abundant mesic and hygrophilous insect fauna in this region before, during, and after the last glacial interval (Elias, 1992c). The faunas contrast sharply with late Quaternary fossil insect assemblages described from interior and northern sites in eastern Beringia, where xeric and steppe-tundra species were dominant. Insect fossils from the Nushagak lowland (inset in Figure 10.1, Nos. 7, 9, 10, 12, 17, and 23) are discussed in Lea et al. (1991). A sampling of photographs of the Nushagak fossils is shown in Fig. 10.3. These assemblages document Wisconsin and Holocene environments. A faunal assemblage from the Nuyakuk River (Fig. 10.1, No. 27) reflects Sangamon (last interglacial) environments (Elias and Short, 1992). That fauna is indicative of climatic conditions at least as warm as those at present, based on the modern distributions of the species in the fossil fauna. Insect faunas from the Holitna and Upper Kuskokwim lowlands (Fig. 10.1, Nos. 14, 15, 21, and 22) are correlated with pre-Sangamon through Holocene environments.

I described only one pre-Sangamon assemblage (Elias, 1992c) from the Kulukbuk Bluffs site that yielded a small assemblage stratigraphically below the last interglacial fauna. This assemblage includes mesic and hygrophilous beetles, a few xeric species, and the rove beetle *Tachinus brevipennis.* This species was an important element in both early and late Wisconsin faunas in southwestern Alaska. In fact, it was practically the only species preserved in many glacial-age assemblages. It is not a xerophilous species but lives only along the northern and western coasts of Alaska, in the Brooks Range, and in eastern Siberia (Campbell, 1988).

Insect faunal assemblages of Sangamon age have been described from the Nuyakuk site (Elias and Short, 1992) and from the Kulukbuk Bluffs site (Elias, 1992c). These two assemblages are characterized by substantial numbers of hygrophilous species and by very low numbers of xeric taxa (Fig. 10.4). The most similar faunal assemblages are of Holocene age. Only the Sangamon and

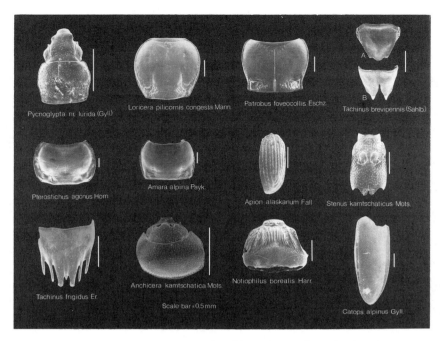

Figure 10.3. Scanning electron micrographs of fossil beetles from sites in southwestern Alaska. Scale bars equal 0.5 mm. (From Lea et al., 1991. Copyright © 1991 by the Regents of the University of Colorado.)

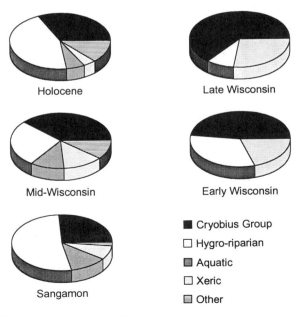

Holocene

Late Wisconsin

Mid-Wisconsin

Early Wisconsin

Sangamon

■ Cryobius Group
☐ Hygro-riparian
▨ Aquatic
☐ Xeric
▤ Other

Figure 10.4. Percentage composition of ecological groups in late Quaternary fossil insect assemblages from southwestern Alaska. (After Elias, 1993.)

Holocene faunas contain bark beetles (Scolytidae), indicative of the proximity of spruce trees to the study sites. Some species of bark beetles feed only on specific host tree species or genera. In this case, the bark beetles attack spruces. The faunas indicate that coniferous forest was established in the Holitna lowland but not in the Nushagak lowland (Elias and Short, 1992).

The few fossil assemblages of early Wisconsin age were recovered from the Nushagak lowland. These are depauperate samples, with fewer hygrophilous and riparian species, no aquatic species, and increased numbers of xeric taxa (Fig. 10.4). The xeric-adapted weevil *Lepidophorus lineaticollis,* abundant in most interior and northern assemblages from eastern Beringia, is found in southwestern Alaska only in this stadial interval. However, unlike the interior faunas, mesic tundra taxa (especially the *Cryobius* group) persisted in southwestern Alaska, even during stadial intervals.

The climatic regimes of the mid-Wisconsin Interstadial were apparently quite varied in southwestern Alaska (Lea et al., 1991), as indicated by the succession of fossil beetle faunas. During the early (and perhaps warmest) part of the interval, the composition of the beetle faunas (Fig. 10.4) was similar to that of the Sangamon and Holocene faunas, except that bark beetle remains and other evidence for coniferous forest are lacking. The hygrophilous element is also not as pronounced as in the full-interglacial faunas, and the xeric element is slightly more prominent. Nevertheless, abundant, diverse insect faunas, rich in mesic tundra species, were preserved in deposits of mid-Wisconsin age. Unfortunately the mid-Wisconsin Interstadial is not dated accurately because much of it is beyond the reliable range of radiocarbon dating. Therefore the rates of environmental change at the onset and within the interstadial cannot be postulated clearly.

Younger (stratigraphically higher) sediments of mid-Wisconsin age yield faunas subtly different from earlier interstadial faunas. The only remarkable difference is that the hygrophilous and riparian species are diminished, compared to earlier mid-Wisconsin faunas. Climatic deterioration is suggested by the presence of the cold-adapted ground beetles *Dyschirius frigidus* and *Patrobus septentrionis* and the rove beetle *Tachinus frigidus,* which are found only in late mid-Wisconsin assemblages.

The last glacial episode brought periglacial environments to the lowlands of southwestern Alaska (Lea, 1989; Waythomas, 1990). Organic deposits were limited to thin layers of detritus dispersed in aeolian silt. Regional beetle faunas were once again reduced to a few species, with most assemblages dominated by *Tachinus brevipennis.* Hygrophilous and riparian taxa were reduced to their lowest levels in the faunal sequence, and aquatic species disappeared from regional records (Fig. 10.4). Xeric taxa increased to their highest levels, even though mesic tundra species continued to be regionally important. The characteristic stadial steppe-tundra fauna from interior and northern sites (*Lepidophorus lineaticollis, Morychus,* and *Harpalus amputatus*) are missing from southwestern assemblages, even during the height of the last glacial episode.

The climatic amelioration that followed the last glacial interval began about 12,500 yr B.P. (Short et al., 1992). Late glacial faunas document this environmental change in several ways. Xeric taxa decline as hygrophilous and riparian species increase in numbers. Furthermore, several thermophilous species are recorded in late glacial deposits, including the ground beetle species *Patrobus stygicus* and the rove beetle species *Tachyporus rulomus.* These species are associated with open ground habitats within the boreal forest. They indicate a boreal climatic regime, even though the boreal forest was still absent from southwestern Alaska.

Holocene assemblages indicate a return of the hygrophilous and riparian species to very high percentages concurrent with a sharp decline in xeric beetles (Fig. 10.4). The coniferous bark beetles *Phloeotribus lecontei* and *Polygraphus rufipennis* were found in Holocene assemblages from the Holitna lowland, but not the Nushagak lowland. The insect evidence, combined with conifer macrofossils, indicates the return of coniferous forest to the former region.

The proximity of southwestern Alaska to maritime moisture probably played an important role in the development of a refugium for mesic and hygrophilous species by maintaining wet habitats even through the last glaciation.

Research in Southeastern Alaska

The only study to date of Quaternary insects from the Alaskan panhandle is my work on Holocene assemblages from Kruzof Island (Fig. 10.1, No. 20). Fossil insect data helped infer Holocene climatic conditions and the development of *Sphagnum* bogs on the island (Klinger et al., 1990). A deciduous woodland covered Kruzof Island from 9000 to 6000 yr B.P., followed by coniferous woodland. About 3000 yr B.P., *Sphagnum* peatland is inferred to have choked the conifer forest, and it has dominated the landscape ever since. The Holocene beetle fauna from the Kruzof peats was comprised mostly of Pacific Northwestern species, rather than taxa typical of eastern Beringian Pleistocene assemblages. This region merits considerably more attention from Quaternary paleoecologists.

This review of Beringian studies has shown that this region was not a uniform, coherent steppe-tundra ecosystem during the late Quaternary. Given the continental size of Beringia, and the variety of environmental conditions impinging on the various landscapes, this should surprise no one. The data in hand represent the tip of a very large iceberg, offering tantalizing glimpses into this vanished realm.

11

OTHER STUDIES IN THE NEW WORLD

The state of our knowledge of the effects of the Pleistocene on the present patterns of insect distribution is, at best, inadequate.

—Henry F. Howden (1969)

In this chapter, I review Quaternary fossil insect studies from the conterminous United States, southern Canada, and South America. Some of these regions have received considerable attention from Quaternary entomologists, whereas many others have not been studied at all. The environmental reconstructions based on insect data are therefore concentrated in a few study regions.

NORTH AMERICAN STUDIES

Quaternary entomology has come a long way in North America since Henry Howden made the foregoing observation. Up to the 1960s, almost the only published Quaternary insect studies were those by Scudder and Pierce, which were fraught with identification errors. During the 1970s, several paleoentomologists turned their attention to North America. By 1980, Morgan and Morgan cited 23 publications treating 33 sites in North America. The number of publications concerning Quaternary entomology has more than tripled in the last ten years. Table 11.1 lists 83 published studies covering 119 sites, in addition to 21 publications for 30 sites in Alaska and the Yukon listed in Table 10.2. Clearly this line of research has grown significantly. Nevertheless, I must also emphasize that Quaternary entomology is still in a pioneering phase in the New World. Research in most regions has dealt with only one or two intervals of time. In the following review, I provide regional summaries of the work thus far published.

Eastern United States and Canada
More than 40 insect fossil studies have been published from sites south of the Laurentide Ice Sheet in eastern North America (Table 11.1, Figs. 11.1 and 11.2). The fossil assemblages range in age from pre-Sangamon Interglacial to

Table 11.1

North American (other than Alaska and the Yukon) Quaternary insect fossil sites and references

Site	Reference(s)
1. Adams Mill, Indiana	Morgan et al. (1983b), Morgan (1987)
2. Ajo Mountains, Arizona	Hall et al. (1989)
3. Athens, Illinois	Morgan (1987)
4. Au Sable River, Michigan	Morgan et al. (1985)
5. Baby Vulture Den, Texas	Elias and Van Devender (1990)
6. Baie du Basin, Magdalen Island, Quebec	Prest et al. (1976)
7. Barehead Creek, Ontario	Warner et al. (1987)
8. Beales, Ohio	Morgan (1987)
9. Bennett Ranch, Texas	Elias and Van Devender (1992)
10. Bida Cave, Arizona	Elias et al. (1992b)
11. Biggsville, Illinois	Carter (1985)
12. Bishop's Cap, New Mexico	Elias and Van Devender (1992)
13. Bongards, Minnesota	Schwert and Ashworth (1985)
14. Brampton, Ontario	Morgan and Freitag (1982), Morgan (1987)
15. Brookside, Nova Scotia	Mott et al. (1986)
16. Butler Mountains, Arizona	Hall et al. (1988)
17. Cañon de la Fragua, Coahuila, Mexico	Elias et al. (1993)
18. Chaudière Valley, Quebec	Matthews et al. (1987)
19. Cincinnati, Ohio	Lowell et al. (1990)
20. Clarksburg, Ontario	Warner et al. (1988)
21. Clinton, Illinois	Morgan (1987)
22. Conklin Quarry, Iowa	Baker et al. (1986)
23. East Milford, Nova Scotia	Mott et al. (1982)
24. Echoing River, Manitoba	Dredge et al. (1990)
25. Eighteen Mile River, Ontario	Ashworth (1977)
26. Elkader, Iowa	Schwert (1992)
27. Ennadai Lake, Keewatin, Northwest Territories	Elias (1982a)
28. Ernst Tinaja, Texas	Elias and Van Devender (1990)
29. Escalante River Caves, Utah	Elias et al. (1992b)
30. False Cougar Cave, Montana	Elias (1990b)
31. Flamborough, Manitoba	Dredge et al. (1990)
32. Fort Dodge, Iowa	Schwert (1992)
33. Fort Francis, Ontario	Schwert and Bajc (1989)
34. Fra Cristobal Mountains, New Mexico	Elias (1987)
35. Gage Street, Kitchener, Ontario	Schwert et al. (1985)
36. Gardena, Illinois	Morgan (1987)
37. Garfield Heights, Ohio	Coope (1968d)
38. Gervais Formation, Minnesota	Ashworth (1980)

Table 11.1 (*Continued*)

Site	Reference(s)
39. Gods River, Manitoba	Dredge et al. (1990)
40. Goulais River, Ontario	Warner et al. (1987)
41. Henday, Manitoba	Nielsen et al. (1986)
42. Hippa Lake, Queen Charlotte Islands, British Columbia	Walker and Mathewes (1988)
43. Hornaday Mountains, Sonora, Mexico	Hall et al. (1988)
44. Hueco Mountains, Texas	Elias and Van Devender (1992)
45. Huntington Canyon, Utah	Elias (1990b)
46. Innerkip, Ontario	Pilny and Morgan (1987)
47. Kaetan Cave, Arizona	Elias et al. (1992b)
48. Kewaunee, Wisconsin	Garry et al. (1990a)
49. Klondike Bog, Mackenzie, Northwest Territories	Matthews (1980b)
50. Lake Emma, Colorado	Elias et al. (1991)
51. Lake Isabelle, Colorado	Elias (1985)
52. Lamb Spring, Colorado	Elias and Nelson (1989), Elias and Toolin (1989)
53. La Poudre Pass, Colorado	Elias (1983), Elias et al. (1986)
54. Last Chance Canyon, New Mexico	Elias and Van Devender (1992)
55. Lefthand Reservoir, Colorado	Elias (1985)
56. Limestone River, Manitoba	Dredge et al. (1990)
57. Lockport Gulf, New York	Miller and Morgan (1982)
58. Longswamp, Pennsylvania	Morgan et al. (1982), Williams et al. (1993)
59. McKittrick, California	Miller (1983)
60. Makinson Inlet, Ellesmere Island, Northwest Territories	Blake and Matthews (1979)
61. Maravillas Canyon, Texas	Elias and Van Devender (1990)
62. Marias Pass, Montana	Elias (1988a)
63. Marion Lake, British Columbia	Walker and Mathewes (1987)
64. Mary Hill, British Columbia	Miller et al. (1985a)
65. Mary Jane, Colorado	Short and Elias (1987)
66. Mike Lake, British Columbia	Walker and Mathewes (1989)
67. Misty Lake, British Columbia	Walker and Mathewes (1989)
68. Mosbeck, Minnesota	Ashworth et al. (1972)
69. Mount Ida Bog, Colorado	Elias (1985)
70. Newton, Pennsylvania	Barnowsky et al. (1988)
71. Nichols Brook, New York	Fritz et al. (1987)
72. Norwood, Minnesota	Ashworth et al. (1981)
73. Oregon Jack Creek, British Columbia	Hebda et al. (1990)

Table 11.1 (*Continued*)

Site	Reference(s)
74. Owl Creek, Ontario	Miller (1990), Mott and DiLabio (1990)
75. Owl Roost, Grand Canyon, Arizona	Elias et al. (1992b)
76. Pasley River, Keewatin, Northwest Territories	Dyke and Matthews (1987)
77. Pointe-Fortune, Quebec	Anderson et al. (1990a)
78. Portage du Cap, Magdalen Island, Quebec	Prest et al. (1976)
79. Portland, Maine	Anderson et al. (1990b)
80. Port Moody, British Columbia	Miller et al. (1985a)
81. Powers, Michigan	Morgan (1987)
82. Puerto Blanco Mountains, Arizona	Hall et al. (1990)
83. Puerto de Ventanillas, Coahuila, Mexico	Elias et al. (1993)
84. Quillin, Ohio	Morgan (1987)
85. Quitman Mountains, Texas	Elias and Van Devender (1992)
86. Rancho La Brea, California	Miller (1983)
87. Roaring River, Colorado	Elias et al. (1986)
88. Rocky Arroyo, New Mexico	Elias and Van Devender (1992)
89. Rostock, Ontario	Pilny et al. (1987)
90. Rous Lake, Ontario	Morgan (1987)
91. Russellville, Indiana	Morgan et al. (1983b), Morgan (1987)
92. Sacramento Mountains, New Mexico	Elias and Van Devender (1992)
93. Salt Creek Canyon, Utah	Elias et al. (1992b)
94. San Andres Mountains, New Mexico	Elias (1987)
95. Sandy River, Maine	Nelson (1987)
96. Saylorville, Iowa	Schwert (1992)
97. Scarborough Bluffs, Ontario	Williams and Morgan (1977), Williams et al. (1981)
98. Seattle, Washington	Nelson and Coope (1982)
99. Seibold, North Dakota	Ashworth and Brophy (1972)
100. Shafter, Texas	Elias and Van Devender (1992)
101. Shelter Cave, New Mexico	Elias and Van Devender (1992)
102. Sierra de Cubabi, Sonora, Mexico	Hall et al. (1988)
103. Sierra de la Misericordia, Durango, Mexico	Elias et al. (1993)
104. Sierra del Rosario, Sonora	Hall et al. (1988)
105. Splan, Portey, and Wood's ponds, New Brunswick	Walker and Paterson (1983), Walker et al. (1991a)
106. St. Charles, Iowa	Baker et al. (1991)
107. Ste. Eugene, Quebec	Mott et al. (1981)
108. Ste. Hilaire, Quebec	Mott et al. (1981)
109. Streeruwitz Hills, Texas	Elias and Van Devender (1992)

Table 11.1 (*Continued*)

Site	Reference(s)
110. Terlingua, Texas	Elias and Van Devender (1990)
111. Tinajas Atlas Mountains, Arizona	Hall et al. (1988)
112. Tonica, Illinois	Schwert (1992)
113. Tunnel View Site, Texas	Elias and Van Devender (1990)
114. Two Creeks, Wisconsin	Morgan and Morgan (1979)
115. Umiakoviarusek, Labrador	Elias (1982c)
116. Wales site, Ontario	Miller et al. (1987)
117. Weaver Drain, Michigan	Morgan et al. (1981), Williams et al. (1993)
118. Wedron, Illinois	Garry et al. (1990b)
119. Winter Gulf, New York	Schwert and Morgan (1980)

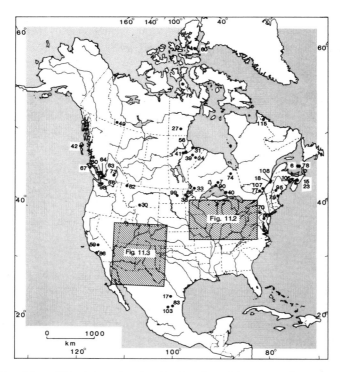

Figure 11.1. Map of North America showing location of Quaternary insect fossil sites. Site numbers are keyed to the list of sites in in Table 11.1.

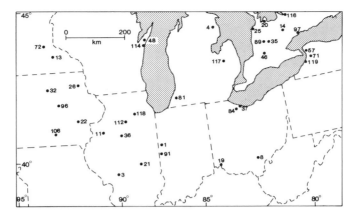

Figure 11.2. Map of eastern North America showing location of regional Quaternary insect fossil sites. Site numbers are keyed to the list of sites in Table 11.1.

late Holocene, but most are of late Wisconsin to early Holocene age. Most Sangamon and Holocene faunas reflect similar environmental conditions, but the composition of regional faunas has changed appreciably between these interglacials. Many of the beetles in the Sangamon faunas are found only in western North America today.

Mott (1990) and Mott and Matthews (1990) reviewed paleoenvironmental evidence from the Sangamon Interglacial in Canada. Because the Sangamon is beyond the range of radiocarbon dating, most assemblages cannot be confidently assigned to strict chronostratigraphic units. Sangamon sites in the Maritime Provinces include a few that contain studied insect assemblages. These are the East Milford site, Nova Scotia (Fig. 11.1, No. 23) (Mott et al., 1982), and the Baie du Basin and Portage du Cap sites (Fig. 11.1, Nos. 6 and 78) on the Magdalen Islands, Quebec (Prest et al., 1976). The faunas indicate warmer than modern conditions, including some species with modern distributions ranging north only to southernmost Canada.

Anderson et al. (1990a) described a Sangamon flora and fauna from the Pointe-Fortune site in southern Quebec (Fig. 11.1, No. 77). As in the Maritimes, these Sangamon assemblages were indicative of warmer than modern climates. The Sangamon deposit was overlain by younger organic horizons that contained a beetle fauna suggesting climatic cooling and hence a waning of interglacial warmth.

Three regions in Ontario have yielded Sangamon insect faunas. These are the Scarborough Bluffs site, Toronto (Fig. 11.2, No. 97) (Williams and Morgan, 1977; Williams et al., 1981); the Innerkip site, southwest of Toronto (Fig. 11.2,

No. 46) (Pilny and Morgan, 1987); and the Owl Creek beds, in northeastern Ontario (Fig. 11.1, No. 74) (Miller, 1990; Mott and DiLabio, 1990). The Scarborough insects are indicative of a climate like that in the region of the lower Great Lakes today, which implies Sangamon temperatures warmer than modern. The Innerkip assemblages are similar in composition to the Scarborough Bluffs assemblages.

The Owl Creek fauna comes from the upper parts of cores from the Timmins region. The fauna is composed of species found today in the boreal and subarctic regions of Canada. It is indicative of northern treeline conditions, a finding that suggests conditions similar to those at present (Miller, 1990). Older sediments in the Owl Creek beds lack insect fossils but have produced pollen spectra suggesting interglacial conditions warmer than those at present (Mott and DiLabio, 1990).

Insect fossils from the Missanaibi Formation in northern Manitoba also indicate interglacial environments. Lists of taxa have been published for five sites, as summarized in Nielsen et al. (1986) and Dredge et al. (1990). Assemblages from the Henday site (Fig. 11.1, No. 41) are indicative of a northern treeline environment, similar to that found nearby today at Churchill, Manitoba. Fossil beetles from the Limestone River site (Fig. 11.1, No. 56) are indicative of open ground environments at or near treeline. These colder than present assemblages may represent the cooler stages of the Sangamon (i.e., isotope stage 5a or 5c). Insects from the Echoing River, Flamborough, and Gods River interglacial deposits (Fig. 11.1, Nos. 24, 31, and 39) reflect boreal zone climate, similar to modern regional conditions. These fossils were probably deposited during the height of Sangamon warming (isotope stage 5e).

Early Wisconsin–age organic sediments, assigned to the Massawippi Formation, have been studied from the Chaudière Valley in southern Quebec (Fig. 11.1, No. 18) (Matthews et al., 1987). Macrofossils from this formation yield infinite radiocarbon ages but are thought to date from a cold stage during the mid-Wisconsin and late Sangamon (isotope stage 3 or 5a), based on their stratigraphic position. The biotic evidence from these deposits suggests the type of arctic environment found today in northernmost Quebec. The insect assemblages include several arctic species that were subsequently extirpated from regions east of Hudson Bay and have not become reestablished in northern Quebec during the Holocene.

Arctic conditions in central Ontario have been demonstrated by an insect assemblage of early- or mid-Wisconsin age from Clarksburg, Ontario (Fig. 11.2, No. 20). The fauna includes the *Pterostichus* (*Cryobius*) species *P. pinguedineus, P. ventricosus,* and *P. brevicornis.* No forest-associated species were found, and mean July temperatures were apparently less than 10°C (Warner et al., 1988).

Coope (1968d) described an insect fauna from Garfield Heights, Ohio (Fig. 11.2, No. 37) that is indicative of conditions before the last Wisconsin glacial advance. This open ground fauna dates from 28,000 to 24,000 yr B.P. No strict tundra dwellers were identified, and isolated conifer macrofossils in the deposit suggest small stands of trees. The identified beetles live today in open ground habitats in eastern Canada. Coope estimated mean July temperatures at the site to have been less than 15°C.

Other contemporaneous sites south of the Great Lakes also had insect faunas suggestive of boreal-style climate and coniferous woodlands. At Athens and Gardena, Illinois (Fig. 11.2, Nos. 3 and 36), Morgan (1987) described insect assemblages ranging in age from 27,000 to 22,500 yr B.P. These are boreal faunas, with species found today in central Canada, where modern July temperatures are 15–17°C.

The Adams Mill and Russellville sites in Indiana (Fig. 11.2, Nos. 1 and 91) date to about 22,000 and 21,000 yr B.P., respectively (Morgan et al., 1983b). These sites yielded small insect faunas, including aquatic beetles, caddisfly larvae, and some scolytids and conifer macrofossils, suggesting the presence of some trees. Mean July temperatures during this interval in Indiana were estimated at 14–16°C.

During the last (Wisconsin) glacial advance, the coniferous woodland ecosystem remained intact only farther to the south, and a band of open ground tundra developed in close proximity to the ice margins. This narrow band of periglacial habitat supported insects with arctic affinity, with mixtures of species not seen in any one region today. Insect fossil sites dating to the last (Wisconsin) glaciation have been described from Clinton, Illinois (Fig. 11.2, No. 21) and from an upper horizon at Gardena (Fig. 11.2, No. 36). The deposits are dated at 20,500 and 19,500 yr B.P., respectively. The older fauna contains a mixture of arctic and boreoarctic beetles. The younger sample includes only arctic and subarctic species. This assemblage reflects mean July temperatures of only 11–12°C (Morgan, 1987).

Another full-glacial fauna has been described by Morgan and Pilny (in Lowell et al., 1990) from Cincinnati, Ohio (Fig. 11.2, No. 19). This deposit has been dated at 19,500 yr B.P. Laurentide ice advanced to within 5 km of the site at 19,600 yr B.P. The beetle fauna suggests mean July temperatures of 10–12°C. These conditions are found today at northern treeline in Canada.

Late glacial insect faunas are suggestive of regional warming after 15,000 yr B.P., but local environmental histories were complicated by waxing and waning of ice lobes and proglacial lakes (see Morgan, 1987). Accordingly, regional late glacial insect faunas reflect a variety of conditions, ranging from arctic in ice-proximal sites to thermophilous in more southerly locations. The earliest of the regional late glacial insect fossil records comes from Longswamp, Pennsylvania (Fig. 11.1, No. 58). These date from 15,000–14,000 yr B.P. Paleobotanical evidence (Watts, 1979) suggests tundra conditions at this time, but even the earliest insect assemblages include boreal species from open ground habitats. These are indicative of mean July temperatures in the 13–14°C range (Morgan et al., 1982; Williams et al., 1993). By 12,500 yr B.P., conifers and bark beetles had arrived. This timing represents a lag of as much as 2500 years between the establishment of boreal zone climate and the arrival of conifers, even though conifer forests were not far south of the site in the late Wisconsin (Watts, 1979).

By 14,500 yr B.P., a boreal climatic regime was established once again in regions immediately south of the Laurentide ice sheet. This fact is demonstrated by an insect fauna from Quillin, Ohio (Fig. 11.2, No. 84), which included boreal insect taxa and conifer macrofossils. Morgan (1987) estimated mean July temperatures of about 15°C for this assemblage. As the ice retreated northward, newly exposed ground supported a mixture of arctic and northern boreal insects, as shown from an assemblage at Weaver Drain, Michigan (Fig. 11.2, No. 117), dated at 13,770 yr B.P. This fauna suggests conditions similar to those at modern treeline in Canada, which equate with mean July temperatures of about 11°C. Although this is considerably cooler than the conditions suggested by the Quillin and Longswamp faunas, it probably reflects persistent regional cooling due to the proximity of the waning Laurentide ice sheet (Morgan et al., 1981). By comparison, a contemporaneous fauna that lived far from the ice margin at Beales, Ohio (Fig. 11.2, No. 8), indicates summer temperatures of 16–17°C (Morgan, 1987).

The late glacial insect faunas from southern Ontario trace the retreat of Laurentide ice followed by rapid warming. Several centuries later, the mixed forest compatible with this warming finally became established. Late glacial assemblages from basal organic sediments in southern Ontario reflect cold, ice-proximal environments. These include faunas from the Rostock site (Fig. 11.2, No. 89) (Pilny et al., 1987), dated at 13,000 yr B.P., and the Brampton site (Fig. 11.2, No. 14) (Morgan and Freitag, 1982; Morgan, 1987), dating to >12,500 yr B.P. at the base. Both basal assemblages contained open ground faunas with no conifer-associated taxa. These ice-marginal faunas are indicative of mean July temperatures of 12–14°C. However, some thermophilous species from Brampton, including the tiger beetle species *Cicindela limbalis,* suggest mean July temperatures of 15°C. Following regional deglaciation, younger assemblages from Rostock (12,000 yr B.P.) indicate mean summer temperatures of 16–18°C, and by 10,000 yr B.P. the Brampton assemblages suggest mean July temperatures rising to 21°C.

The most complete late glacial insect sequence is from the Gage Street site, in Kitchener, Ontario (Fig. 11.2, No. 35) (Schwert et al., 1985). The base of the organic sequence was dated at 12,800 yr B.P. and contained a mixture of open ground and boreal species, including bark beetles. Mean July temperatures reflected by this assemblage were 15–17°C at 12,800, rising to nearly 21°C by 10,000 yr B.P. The Holocene fauna (8600–7900 yr B.P.) indicates summer temperatures of 22–23°C, followed by an increase to 25°C by 6900 yr B.P. at the top of the sedimentary sequence. However, sites north of the Great Lakes in Ontario indicate summer temperatures only in the 17–18°C range at this time. These sites include Barehead Creek, north of Lake Superior (Fig. 11.1, No. 7), and the Goulais River site near Sault Ste. Marie (Fig. 11.1, No. 40) (Warner et al., 1987).

Late glacial sites in upstate New York were apparently not affected by stagnant ice or cold meltwater from the proto–Great Lakes. Assemblages from Winter Gulf (Fig. 11.2, No. 119) date to 12,700–12,500 yr B.P. The beetle fauna implies an open mire surrounded by trees, and mean July temperatures of greater than 16°C (Schwert and Morgan, 1980). A late glacial site at Nichols Brook, New York (Fig. 11.2, No. 71), reported by Fritz et al. (1987), is contemporaneous with Winter Gulf. The basal insect fauna from Nichols Brook is indicative of conditions at the center of the boreal zone. Younger assemblages show that summer temperatures rose to 18°C by 11,000 yr B.P., and perhaps to 20°C by 10,000 yr B.P.

Ashworth and Schwert (in Barnowsky et al., 1988) published the results of a study of late glacial insects associated with a mammoth skeleton at Newton, Pennsylvania (Fig. 11.1, No. 70). The insect assemblages were from a horizon dated at 12,100 yr B.P. and from below this horizon. The insects represent open ground conditions and climate similar to that at modern treeline. The assemblages include *Pterostichus pinguedineus* and *Diacheila polita;* the latter species is thought to have been extirpated from the American Midwest by about 15,000 yr B.P.

Another fauna that lived far from the Laurentide ice margin was described from Two Creeks, Wisconsin (Fig. 11.2, No. 114) (Morgan and Morgan, 1979). Assemblages dating from the interval 12,000–11,800 yr B.P. comprise an open boreal fauna reflecting July temperatures of 15–16°C.

Garry et al. (1990a) described a contemporaneous beetle fauna from Kewaunee, Wisconsin (Fig. 11.2, No. 48). Although the Kewaunee fauna has many species in common with that at Two Creeks, it also contains some cold stenotherms indicative of slightly colder conditions. The difference in temperature reconstruction may reflect microclimatic differences between the sites.

Anderson et al. (1990b) studied late glacial sediments from Portland, Maine (Fig. 11.1, No. 79). Insect fossils dating to 11,500 yr B.P. included boreal zone beetles and ants, but some of the beetles are indicative of open ground situations within the boreal forest.

The Champlain Sea was a large proglacial lake, occupying much of southeastern Ontario, southwestern Quebec, and adjacent regions of New England. Insect fossil assemblages have been described from sites along the Champlain Sea at Ste. Eugene and Ste. Hilaire, Quebec (Mott et al., 1981). The Ste. Eugene site (Fig. 11.1, No. 107), dated at 11,050 yr B.P., includes a northern boreal and treeline beetle assemblage, indicative of July temperatures averaging 17–18°C. The Ste. Hilaire site (Fig. 11.1, No. 108), dated at 10,100 yr B.P., contains thermophilous insects indicating mean July temperatures of 19–20°C.

Late glacial boreal forest environments have also been documented in Michigan and New York. At the Powers site in southern Michigan (Fig. 11.2, No. 81), a boreal fauna was found in association with mastodon remains dated at 11,200 yr B.P. This fauna reflects mean July temperatures of 17°C (Morgan, 1987). Miller and Morgan (1982) studied insect fossils from contemporaneous assemblages at Lockport Gulf, New York (Fig. 11.2, No. 57). Organic remains from a pond deposit spanning 10,900–9100 yr B.P. yielded abundant insect assemblages. The faunas are indicative of boreal forest habitats from 10,900 to 9800, followed by a transition to a mixed forest fauna from 9700 to 9100 yr B.P. As inferred from many other insect faunas in the late glacial to Holocene transition, the Lockport Gulf faunas indicate that late glacial climatic conditions were already suitable for the establishment of the plant communities that would finally become stable in the Holocene. The vegetation succession from boreal to mixed forest therefore reflects ecological succession rather than climatic warming in the early Holocene.

At Eighteen Mile River, Ontario (Fig. 11.2, No. 25), a site dated at 10,600 yr B.P. documented conditions on the margin of glacial Lake Algonquin (Ashworth, 1977). This fauna contained a mixture of elements (mostly boreal) with no modern analogue. As at the Kewaunee site in Wisconsin, the Eighteen Mile River assemblages suggest a cold microenvironment surrounded by a warm macroclimate. Mean July temperatures at the site were as low as 12–13°C, whereas other regional sites at this time show summer temperatures of 18–19°C.

In the Canadian Maritimes, Mott et al. (1986) studied a late glacial site near Brookside, Nova Scotia (Fig. 11.1, No. 15). Organic deposits there ranged in age from 11,700 to 11,000 yr B.P. The insect faunal assemblages were indicative of northern treeline conditions, although the pollen spectra suggested shrub tundra. Organic deposition ceased shortly after 11,000 yr B.P., and Mott et al. attributed this finding to an abrupt shift to colder climatic conditions, equivalent to the Younger Dryas cooling in northwestern Europe.

Fossil chironomid (midge fly) larval faunas are beginning to provide some very interesting paleoenvironmental data for North American sites. Synchronous abrupt climatic deterioration has been inferred from chironomid larval assemblages from sediments in Splan Pond, New Brunswick (Fig. 11.1, No. 105) (Walker et al., 1991a). Pollen evidence from a suite of sites in the Canadian Maritimes also indicates an abrupt climatic reversal at about 11,000 yr B.P. However, fossil insect assemblages from regions west of the Canadian Maritime Provinces have not indicated this climatic reversal.

Walker and Paterson (1983) identified Holocene chironomid faunas from Portey and Wood's ponds in New Brunswick (Fig. 11.1, No. 105). These studies suggested changes in the trophic status of the ponds and highlighted the differences between ponds that become choked with *Sphagnum* mosses and those that remain free of *Sphagnum.*

A mid-Holocene age exposure on the Au Sable River, Michigan (Fig. 11.2, No. 4) (Morgan et al., 1985), yielded essentially a modern beetle fauna, with species found today in northern and central Michigan. At this site, the paleoenvironmental reconstructions based on insects and plants reached total agreement only by 4000 yr B.P.

A late Holocene fauna from the Sandy River, Maine (Fig. 11.1, No. 95), likewise yielded insects in harmony with the paleobotanical evidence (Nelson, 1987). This assemblage dates to 2100 yr B.P.

Midwestern Studies

The oldest Quaternary insect fauna in the midwestern region is beyond the range of radiocarbon dating but is probably early Wisconsin in age. It is an assemblage from the Gervais Formation in northwestern Minnesota (Fig. 11.1, No. 38) (Ashworth, 1980). The fauna is a mixture of boreal and arctic taxa, reflecting conditions similar to those at modern treeline in Canada. Some of the species are known today only from northwestern North America, including the weevil *Vitavitus thulius,* which is common in interstadial assemblages from eastern Beringia.

The other insect faunas from this region range in age from 34,000 to 11,700 yr B.P. They have been studied from sites in Illinois, Iowa, Minnesota, North Dakota, Wisconsin, and northwestern Ontario. The fossil faunas indicate regional responses to a series of marked environmental changes from before the last glaciation through the present. The earliest of these faunas is from the St. Charles site in Iowa (Fig. 11.2, No. 106) (Baker et al., 1991). Insect assemblages dated at 34,000 yr B.P. are indicative of open ground environments, but all species live today in the upper Great Lakes region, which suggests that the climate was only sightly cooler than that of modern times.

The midwestern faunal sequence in the late Wisconsin is similar to that discussed earlier for regions farther east. Midwestern regional sites that predate the last glaciation span the interval 27,000–21,000 yr B.P. These are indicative of closed spruce forest. Boreal faunas were replaced by tundra and forest-tundra faunas during the late Wisconsin glaciation. By 14,000 yr B.P. the cold-adapted species were extirpated by climatic warming and replaced by a fauna of boreal and western montane species. This paleoenvironmental reconstruction has recently been summarized by Schwert (1992). The results of individual studies are summarized in Table 11.2.

Sites predating the last glaciation include Biggsville, Illinois (Fig. 11.2, No. 11), Gardena, Illinois (Fig. 11.2, No. 36), Athens, Illinois (Fig. 11.2, No. 3), and Wedron, Illinois (Fig. 11.2, No. 118). These assemblages are characterized by diverse forest faunas, dominated by scolytids and weevils, and swamp-dwelling carabids and staphylinids. Close modern analogues for these assemblages are to be found in the southern boreal forest regions of central Canada. By 23,000 B.P. the forest had been replaced by open ground as glaciers spread southward.

Assemblages from the last glaciation have been described from the upper section of the Gardena site, Conklin Quarry, Iowa (Fig. 11.2, No. 22), Saylorville, Iowa (Fig. 11.2, No. 96), and Fort Dodge, Iowa (Fig. 11.2, No. 32). These faunas are characterized by subarctic species, with modern analogues in tree-line situations west of the Mackenzie Delta region of northwestern Canada. Mean July temperatures in this region are 10–12°C.

Late glacial sites in the midwestern United States are also summarized in Table 11.2. These include the upper part of the Ft. Dodge sequence; Tonica, Illinois (Fig. 11.2, No. 112); upper horizons of the site at Saylorville, Iowa; and Norwood, Minnesota (Fig. 11.2, No. 72). Late glacial faunas are dominated by subarctic species until 14,000 B.P. at Saylorville but contain a mixture of open ground species with no exact modern analogues. By 14,300 yr B.P. at Tonica, and by 12,400 yr B.P. at Norwood, the open ground fauna had been replaced by closed spruce forest species, even though the sites bordered the ice margin. As ice retreated, newly exposed ground was colonized by thermophilous, open ground species; these were rapidly replaced by boreal forest species. Holocene sites in the midwestern United States are also summarized in Table 11.2. These include Seibold, North Dakota (Fig. 11.1, No. 99), and Bongards, Minnesota (Fig. 11.2, No. 13).

One of the major trends in the midwestern scenario was the regional extinction of arctic species at the end of the last glaciation. These beetles had nowhere to go in the midwest because their migration northward was blocked by stagnant ice, and local conditions were becoming increasingly too warm for

Table 11.2
Summary of Quaternary insect fossil sites in the midwestern United States

Site	Age range (yr B.P.)	Environmental interpretations	Reference
Pre–last glacial faunas			
Biggsville, Illinois	27,900–21,400	Diverse forest assemblage, similar to modern fauna of southern boreal forest in central Canada.	Carter (1985)
Gardena, Illinois	25,400	Diverse forest assemblage, similar to modern fauna of southern boreal forest in central Canada.	Morgan and Morgan (1986)
Wedron, Illinois	21,500	Diverse forest assemblage, similar to modern fauna of southern boreal forest in central Canada.	Garry et al. (1990b)
Last glacial faunas			
Gardena, Illinois	19,700	Subarctic assemblage with modern analogues near treeline in Mackenzie Delta region of Canadian Northwest Territories; regional mean July temperatures = 10–12°C.	Morgan and Morgan (1986)
Conklin Quarry, Iowa	18,100–16,700	Subarctic assemblage with modern analogues near treeline in Mackenzie Delta region of Canadian Northwest Territories; regional mean July temperatures = 10–12°C.	Baker et al. (1986)

Site	Date (yr B.P.)	Description	Reference
Saylorville, Iowa	17,000–14,000	Subarctic assemblage with modern analogues near treeline in Mackenzie Delta region of Canadian Northwest Territories; regional mean July temperatures = 10–12°C.	Schwert (1992)
Ft. Dodge, Iowa	15,400–15,100	Subarctic assemblage with modern analogues near treeline in Mackenzie Delta region of Canadian Northwest Territories; regional mean July temperatures = 10–12°C.	Schwert (1992)
Late glacial sites			
Ft. Dodge, Iowa	14,400–13,400	Subarctic assemblages with open ground species; no modern analogue.	Schwert (1992)
Tonica, Illinois	14,300	Open ground faunas replaced by spruce forest faunas.	Schwert (1992)
Saylorville, Iowa	14,000–13,100	Open ground faunas replaced by spruce forest faunas.	Schwert (1992)
Norwood, Minnesota	13,000–11,200	Subarctic assemblages replaced by spruce forest faunas after 12,400 yr B.P.	Ashworth et al. (1981)
Ft. Frances, Ontario	11,000–10,000	Fauna from Lake Agassiz shore indicative of boreal forest found today in central and southern Canada.	Schwert and Bajc (1989)
Holocene sites			
Seibold, North Dakota	9800–?	Documents transition from boreal forest to tall grass prairie between 9000 and 8500 yr B.P.	Ashworth and Brophy (1972)
Bongards, Minnesota	3500–recent	Assemblages reflect modern conditions throughout sequence.	Schwert and Ashworth (1985)

these cold stenotherms to tolerate. Some species managed to migrate into the Appalachian Mountains, where they are described as Pleistocene relicts today. Most others were extirpated south of the waning ice sheet; Beringian populations of these species have recolonized western Canada during the Holocene. In eastern Canada, late-lying ice made northern migration almost impossible, because it persisted well into the mid-Holocene (see Chapter 6 for details).

Studies in Arctic and Subarctic Canada

Few fossil insect investigations have been undertaken in northern Canada, split between sites representing the last interglacial and the Holocene. The Laurentide Ice Sheet persisted on the central Labrador-Ungava peninsula of eastern Canada until about 7000 yr B.P. (Dyke and Prest, 1986). This late-lying ice effectively blocked the immigration of insects until the mid-Holocene. I studied a late Holocene insect fauna from the Umiakoviarusek site in northeastern Labrador (Fig. 11.1, No. 115) (Elias, 1982c). The peat profile from which the insects were extracted was dated between 2650 yr B.P. and recent. Assemblages from the interval 2650–1000 yr B.P. were indicative of a boreal woodland environment. Younger assemblages suggested a climatic cooling as arctic tundra species became dominant.

I also worked on Holocene insect fossils from two peat bank profiles at Ennadai Lake, Keewatin, Northwest Territories (Fig. 11.1, No. 27) (Elias, 1982a). The lake is situated at northern treeline today. The Ennadai region was deglaciated by about 8000 yr B.P. (Dyke and Prest, 1986). Peat growth probably commenced after the draining of glacial Lake Kazan, which occupied the region as the ice sheet melted. The Ennadai faunas span the interval 6300–600 yr B.P. The earliest peat layers contain evidence of spruce establishment, but the insects from the basal peat reflect open ground environments, and no bark beetles were found in the basal peats (circa 6300–6000 yr B.P.). Between 6000 and 2800 yr B.P., insect assemblages typical of northern boreal forest were dominant. A decline in conifer pollen from 4800 to 4500 yr B.P. had previously been interpreted as a climatic cooling, forcing a shift in treeline to points south of Ennadai Lake (Nichols, 1975). However, coniferous bark beetles persisted at Ennadai through this interval, documenting the continued presence of trees. I inferred from this finding that regional stands of spruce underwent some type of stress that caused them to reduce or stop pollen production for a few centuries.

Both insect and paleobotanical data indicated a climatic cooling and forest retreat between 2200 and 1500 yr B.P., followed by a return of woodland by about 1000 yr B.P.

Dyke and Matthews (1987) studied organic deposits exposed on the Pasley River, Boothia Peninsula, northern Keewatin (Fig. 11.1, No. 76). These deposits contained plant macrofossils and insects indicative of conditions warmer than those at present and have been tentatively correlated with the Sangamon Interglacial. The provenance of some of the organic horizons is confounded by the inclusion of reworked organic materials, some of which may have come from Beaufort Formation (late Tertiary) deposits.

Blake and Matthews (1979) studied another peat deposit from a site well beyond modern treeline in the Canadian arctic. Peat from Makinson Inlet, Ellesmere Island (Fig. 11.1, No. 59) yielded three infinite radiocarbon ages. A small insect fauna extracted from the peat included only species with modern distributions south of the study site. Blake and Matthews concluded that the deposit was formed during an interval with a warmer than modern climate, probably the last interglacial.

Matthews (1980b) studied insect fossils from John Klondike Bog, in southwestern Mackenzie, Northwest Territories (Fig. 11.1, No. 49). The samples were taken from a core rather than from an exposure. The few species in the sample, initially thought to be of Holocene age, indicated little departure from present climatic conditions. The assemblage is now thought to represent a Wisconsin Interstadial environment (J. V. Matthews, Jr., written communication, 1992).

Studies in the Rocky Mountain Region

The vertical relief of the Rocky Mountains has provided the regional biota with a wide variety of physical environments, expressed in the development of biomes ranging from grasslands in the east through montane and subalpine forests and alpine tundra. Ecotones between these biomes have undergone measurable shifts in elevation in the late Quaternary, facilitating the study of environmental changes. Fossil insects from the Rocky Mountains have left records of environmental change not found in adjacent lowlands.

I have studied a north-south transect of late glacial assemblages from peat bogs, sedge fens, and lake sediments (Elias, 1988b, 1990b, 1993). The Rockies have been a refuge for many cold-adapted species of both plants and animals. Species that shifted southward in front of advancing glacial ice in the American interior had only two options when the ice receded. They had to migrate either northward or up the slopes of mountains. Thus, cold-adapted beetles that lived on the plains south of Denver toward the end of the last glaciation retreated to the alpine tundra in Colorado.

The transect of Rocky Mountain sites is summarized in Table 11.3. Late glacial sites include Marias Pass, Montana (Fig. 11.1, No. 62), False Cougar Cave, Montana (Fig. 11.1, No. 30), Mary Jane, Colorado (Fig. 11.3, No. 65), and Lamb Spring, Colorado (Fig. 11.3, No. 52). Based on the modern distribu-

Table 11.3

Summary of Quaternary insect fossil sites in the Rocky Mountain region of the United States

Site	Age range (yr B.P.)	Environmental interpretations	Reference(s)
Last glacial faunas			
Lamb Spring, Colorado (1731 m elevation)	18,000	Cold-adapted grassland species indicate cold, dry conditions similar to modern grassland regions in western Canada; summer temperatures depressed by about 8–10°C.	Elias and Nelson (1989), Elias and Toolin (1989)
Late glacial sites			
Lamb Spring, Colorado (1731 m elevation)	14,500	Alpine tundra species indicate cold, moist conditions and open ground environment; summer temperatures depressed by about 8–10°C.	Elias and Nelson (1989), Elias and Toolin (1989)
Mary Jane, Colorado (2882 m elevation)	13,740–12,380	Treeline depressed to below site. Alpine tundra fauna suggests summer temperatures depressed by 7°C at 13,700 B.P.; after 12,800 yr B.P., treeline was near site, and summer temperatures were depressed by about 5–6°C.	Short and Elias (1987)
Marias Pass, Montana (1548 m elevation)	12,000	Arctic and alpine fauna indicates treeline depression of at least 650 m below modern level; this equates with a summer temperature depression of at least 5°C.	Elias (1988a)
Holocene sites			
False Cougar Cave, Montana (2590 m elevation)	10,000	Insect fauna suggests essentially modern climatic conditions.	Elias (1990b)

Site	Age (yr B.P.)	Interpretation	Reference
La Poudre Pass, Colorado (3103 m elevation)	9850–Recent	Basal assemblages indicate conditions as warm as modern, although conifer forest arrived 500 yr later. Faunas show climatic cooling from about 6000–5500 and about 4500 to 2700 yr B.P. Ameliorations peaked at 8000, 5000, and 2000 yr B.P.	Elias (1983), Elias et al. (1986)
Huntington Canyon, Utah (2730 m elevation)	9400	Insect fauna suggests essentially modern climatic conditions.	Elias (1990b), Gillette and Madsen (1992)
Lake Isabelle, Colorado (3324 m elevation)	9000–500	Oldest assemblages indicate climatic conditions at least as warm as modern; climatic cooling suggested by faunas dating from 4500 yr B.P. Younger faunas indicate close proximity of treeline from 3800–500 yr B.P.	Elias (1985)
Lake Emma, Colorado (3740 m elevation)	9000–8200	Insect assemblages show treeline and summer temperatures above modern levels.	Elias et al. (1991)
Mount Ida Pond, Colorado (3520 m elevation)	9070–4600	Insects suggest modern climatic conditions from 9000 to 8300 yr B.P. Climatic optimum reached just after 8300 yr B.P; then a gradual cooling is in evidence, with youngest fauna showing coolest conditions.	Elias (1985)
Lefthand Reservoir, Colorado (3224 m elevation)	5350–Recent	Insects suggest cooler than modern climate from 4800 to 3600 yr B.P, then rapid warming at 3000–2800 yr B.P., a cool interval from 2000 to 1400 yr B.P., and a moderate warming about 800 yr B.P.	Elias (1985)
Roaring River, Colorado (2800 m elevation)	2400	Insect fauna indicative of modern conditions.	Elias et al. (1986)

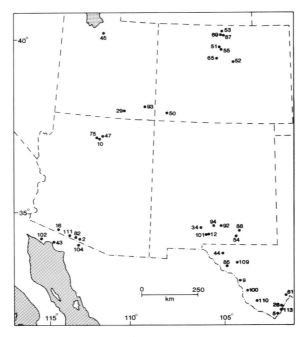

Figure 11.3. Map of southwestern North America showing location of regional Quaternary insect fossil sites. Site numbers are keyed to the list of sites in Table 11.1.

tions and thermal tolerances of insect species in the fossil assemblages from these sites, summer temperatures were depressed by about 8–10°C from 18,000 to 14,000 yr B.P. (Elias, 1991). During this interval, regional insect communities comprised mixtures of species no longer found coexisting in any one region. This level of cooling has also been postulated by glacial geologists as the necessary depression of summer temperature to allow the growth of mountain glaciers to their maximum late Wisconsin (Pinedale) size (Leonard, 1988).

Beginning at about 14,000 yr B.P., regional climates began warming, based on the replacement of cold-adapted beetle species with thermophilous taxa. Fossil insect data suggest that this amelioration was gradual from 14,000 to approximately 11,500 yr B.P. During this interval, insect fossil evidence suggests that summer temperatures rose from about 10°C cooler than present to about 5°C cooler than present. After 11,500 yr B.P., insect fossil data suggest that regional climates warmed extremely rapidly, with summer temperatures reaching modern conditions within a few centuries. Insect faunas indicative of a treeline higher than the present one suggest that early Holocene climates may have been warmer than those at present.

Holocene sites in the Rocky Mountain region are summarized in Table 11.3. These include La Poudre Pass, Colorado (Fig. 11.3, No. 53), Lake Emma, Colorado (Fig. 11.3, No. 50), Roaring River, Colorado (Fig. 11.3, No. 87), Mount Ida Bog, Colorado (Fig. 11.3, No. 69), Lake Isabelle, Colorado (Fig. 11.3, No. 51), Lefthand Reservoir, Colorado (Fig. 11.3, No. 55), and Huntington Canyon, Utah (Fig. 11.3, No. 45).

During the Holocene, the Rocky Mountain region has experienced a series of climatic fluctuations, with insect assemblages indicative of conditions warmer than those at present between 9500 and 7000 yr B.P. and conditions colder than those at present between 4500 and 3000 yr B.P. and again within the last 1000 years. The insect response has essentially been in phase with vegetational changes, but the postulated shift of altitudinal treeline lagged behind the shift in insect faunas by about 500 years.

Pacific Coast Studies

Miller (1983) listed the taxa and provenance of insect fossil assemblages from California asphalt deposits at McKittrick (Fig. 11.1, No. 59) and Rancho La Brea (Fig. 11.1, No. 86). Following Pierce's (1946–1957) misidentifications (see Chapter 1), these fossils have been reevaluated as extant species. The Rancho La Brea assemblages range in age from >40,000 yr B.P. to recent, and the McKittrick fauna dates to about 7000 yr B.P. Little attempt has been made to use these fossils to reconstruct regional paleoenvironmental conditions, although some assemblages have been shown to reflect conditions similar those in southern California today.

Nelson and Coope (1982) described a diverse insect fauna dating from 16,600 yr B.P. from Seattle, Washington (Fig. 11.1, No. 98). Although the pollen spectra associated with this assemblage suggest conditions substantially colder than those at present, the insects are characteristic of the modern Puget lowland. Nelson and Coope suggest that the discrepancy between the flora and insect fauna may be due to increased climatic continentality just before the last (Vashon) glacial advance.

In southern British Columbia, Miller et al. (1985a) studied insect fossils dated from 18,700 to 18,300 yr B.P. at Mary Hill and Port Moody (Fig. 11.1, Nos. 64 and 80). These assemblages represent an open coniferous forest floor community, developed in cool, dry climatic conditions between two advances of the Cordilleran Ice Sheet.

Walker and Mathewes (1988) investigated chironomid larval fossils dated at from 11,000 yr B.P. to recent from sediments at Hippa Lake in the Queen Charlotte Islands (Fig. 11.1, No. 42). Basal sediments yielded chironomids with arctic and alpine affinities. Cold stenotherms persisted in the lake through the Holocene, possibly because of the flow of cold springwater into this shallow lake.

Walker and Mathewes (1989) also studied chironomids from Mike Lake and Misty Lake in southwestern British Columbia (Fig. 11.1, Nos. 66 and 67). Cold stenotherms adapted to oligotrophic waters dominated late glacial sediments but were almost extirpated during early Holocene climatic amelioration.

Hebda et al. (1990) examined insect fossils from two late Holocene packrat middens in the arid interior of British Columbia. The Oregon Jack Creek site (Fig. 11.1, No. 73) yielded a small fauna dated at 1100 yr B.P. This is the northernmost packrat midden locality to produce identifiable insect fossil remains. The Oregon Jack Creek fauna was composed of silken fungus beetles (Cryptophagidae), minute brown scavenger beetles (Lathridiidae), and click beetles (Elateridae). These beetles were probably scavengers in the packrat's nest. Although all these families have been found in packrat midden assemblages from the desert southwest, none is an important constituent there. Conversely, beetle families that dominate desert midden assemblages, such as darkling beetles (Tenebrionidae), spider beetles (Ptinidae), and checkered beetles (Cleridae), were absent from the Oregon Jack Creek assemblages.

Studies in the Desert Southwest

Elias (1987, 1992b), Elias and Van Devender (1990, 1992), and Elias et al. (1993) studied insect fossils from a north-south transect of packrat midden sites in the Chihuahuan Desert. The Chihuahuan Desert is an interior continental desert, stretching from southeastern Arizona to northern Zacatecas, Mexico. Three sites in the Bolson de Mapimi region, in the southern Chihuahuan Desert of Mexico (Fig. 11.3, Nos. 17, 83, and 103), yielded 45 fossil packrat midden assemblages ranging in age from 13,500 yr B.P. to recent. These middens contained insects, arachnids, and millipedes; most of the identified taxa are beetles. Modern studies have shown that the great majority of arthropods found in packrat nests are facultative inquilines. This means that they are species that normally dwell in the open but occasionally enter the rockshelters housing packrat nests in order to take advantage of the moderated microclimate. Once there, many of these arthropods prey on other species or scavenge food from the piles of plant detritus accumulated by the rat.

The fossil assemblages from the Bolson de Mapimi region are unique among those studied thus far from the Chihuahuan Desert because they contain mixtures of desert and temperate zone species in almost every chronozone from the late glacial through the late Holocene. Midden assemblages from locations farther north in the Chihuahuan Desert are generally separated into glacial age faunas with temperate zone affinities and postglacial faunas with desert zone

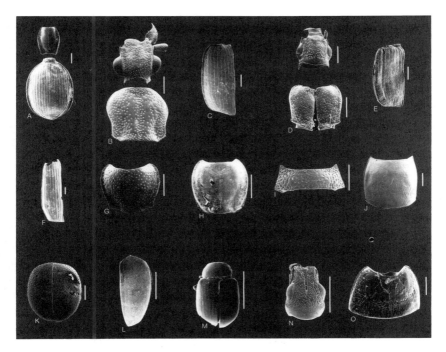

Figure 11.4. Scanning electron micrographs of insect fossils from packrat midden assemblages from the northern Chihuahuan Desert in Texas and New Mexico. A, pronotum and abdomen of *Agonum (Rhadine) longicolle* (Carabidae); B, head and pronotum of *Hellumorphoides texanus* (Carabidae); C, left elytron of *Pinacodera platicollis* (Carabidae); D, head and pronotum of *Cymindis interior* (Carabidae); E, left elytron of *Lebia lecontei* (Carabidae); F, left elytron of *Cymindis arizonensis* (Carabidae); G, pronotum of *C. arizonensis;* H, pronotum of *Synuchus impunctatus* (Carabidae); I, pronotum of *Endrotes rotundatus* (Tenebrionidae); J, pronotum of *Eleodes* cf. *goryi* (Tenebrionidae); K, abdomen of *Endrotes rotundatus;* L, right elytron of *Choleva americana* (Leptodiridae); M, pronotum and abdomen of *Chaetocnema elongulata* (Chrysomelidae); N, head of *Pantomorus elegans* (Curculionidae); O, pronotum of *Pangaeus congruus* (Hemiptera: Cydnidae). Scale bars equal 0.5 mm. (Photographs by the author.)

affinities. The "no modern analogue" faunal assemblages indicate that the late Quaternary environments in this part of the desert were unlike any that exist today (Elias et al., 1993). This conclusion is also borne out in the paleobotanical record from the Bolson de Mapimi, which likewise shows unique combinations of floristic elements.

Elias and Van Devender (1990) also studied fossil insects from 50 assemblages at five localities in and around Big Bend National Park, Texas (Fig. 11.3,

Nos. 5, 28, 61, 110, and 113), the samples ranging in age from >36,000 yr B.P. to recent. Plant macrofossil analyses have been conducted and radiocarbon chronologies have been developed for these and the Chihuahuan Desert sites.

A third part of my study included a transect of 19 midden sites in the northern Chihuahuan Desert of western Texas and southern New Mexico (Fig. 11.3, Nos. 9, 12, 34, 44, 54, 85, 88, 92, 94, 100, 101, and 109). These samples range in age from >43,300 yr B.P. to recent, yielding a wide variety of insects, arachnids, and millipedes (Fig. 11.4).

The late glacial assemblages from the southern Chihuahuan Desert, dated between 13,590 and 12,280 yr B.P., comprise a mixture of temperate and desert taxa. However, most of the desert-associated species are phytophages (plant feeders), more indicative of certain desert floristic elements than of xeric climate per se. The late glacial faunas have species with tropical to subtropical affinities. The only species in the late glacial assemblages that today is confined the temperate zone of the United States is the ground beetle species *Amara chalcea,* which lives in xeric situations within these regions.

A temporal hiatus in samples from 12,000 to 9000 yr B.P. is followed by a loss of most of the late glacial insect species from the southern Chihuahuan records. Beetles associated with desert scrub communities dominated early and middle Holocene assemblages. In the late Holocene, several insect species were recorded, including both temperate and desert species. Paleobotanical and arthropod records indicate that this region served as a refugium for desert biota during the late Pleistocene. Its southerly latitude appears to have dampened the climatic effects of the Wisconsin glaciation, allowing the survival of desert species during the late glacial interval and presumably earlier. A mosaic of temperate and xeric habitats was available to the regional biota, even during the last 1000 yr, when other regions of the Chihuahuan Desert have experienced extremes of aridity.

The Big Bend regional faunas suggest greater effective moisture from 30,000 to 12,000 yr B.P. An especially diverse, mesic fauna characterized the earlier half of this interval (30,000–20,000 yr B.P.). In the late Wisconsin, many grassland species, now confined to higher elevations and cooler, moister regions to the north, lived in the Big Bend region. After 12,000 yr B.P., most of these species were replaced by either desert species or more cosmopolitan taxa. Few stenothermic species were preserved in the late Wisconsin records from Big Bend, but the faunal change suggested a climatic shift from the cool, moist conditions of Wisconsin glacial times to the hotter, drier conditions of latest Wisconsin and early Holocene.

In the northern Chihuahuan Desert, full-glacial (22,000–18,000 yr B.P.) arthropod records suggest widespread coniferous woodland at elevations as low as 1200–1400 m above sea level. From about 22,000 to 11,000 yr B.P., woodland environments persisted, but the insect data suggest considerable open ground, with grasses at least locally important at the midden sites. The grassland nature of the arthropod fauna was also suggested in the regional vertebrate record. The transition from the temperate Wisconsin fauna to the more xeric postglacial fauna started by 12,500 yr B.P., the timing of this faunal change being essentially synchronous throughout the northern Chihuahuan Desert. The Big Bend assemblages suggested that the transition took place after 12,000 yr B.P., but that chronology was based on far fewer data.

A major difference between the Big Bend and northern Chihuahuan Desert scenarios lies in the nature of this temperate-to-xeric faunal change. In the Big Bend region, the transition was characterized by the disappearance from the record of all but one of the temperate insect species at about 12,000 yr B.P. However, the xeric-adapted fauna did not appear in the Big Bend records until about 7500 yr B.P. In the northern Chihuahuan Desert assemblages, the xeric species first appeared at 12,500 yr B.P., and several of the temperate grassland species from the Wisconsin interval persisted well into the Holocene. This mixture of xeric and temperate elements makes sense from an ecological standpoint, given that these northern faunas were living close to the edge of the Chihuahuan Desert. The gradual shifting of northern desert boundaries in the Holocene probably created many marginal habitats for temperate species in ecotones between grassland and desert scrub.

By about 7500 yr B.P., the appearance of more xeric species indicates the establishment of desert environments, including desert grasslands, throughout the Chihuahuan Desert region. After 2500 yr B.P., the last of the temperate species were replaced by species associated with desert scrub communities.

On the whole, there is good agreement between regional reconstructions based on paleobotanical evidence and the fossil arthropod evidence. The two sets of records agree on the nature of Wisconsin-age and Holocene environments, including temperature and moisture conditions. The previous study of insect fossils compared with paleobotanical data from the nearby Fra Cristobal and San Andres Mountains in south-central New Mexico (Fig. 11.3, Nos. 34 and 94) (Elias, 1987) revealed similar agreement. The only serious disparity between insect and plant macrofossil records comes during intervals of rapid climate change, especially the transition from Wisconsin glacial to Holocene (postglacial) environments. In this case, the arthropod evidence shows that temperate Wisconsin environments gave way to postglacial environments with increasing temperatures and shifting precipitation patterns by 12,500 yr B.P. The principal botanical change that marks this transition, the disappearance of piñon pines, occurred between about 11,000 and 10,250 yr B.P., depending on the region. However, detailed analyses of the Hueco Mountains flora revealed shifts in the abundance and diversity of plants, including piñon pine, at about 12,500 yr B.P. Xeric desert scrub species such as lechugilla, creosote bush, and ocotillo appeared in the Chihuahuan Desert paleobotanical records about 4500 yr B.P. Xeric-adapted insects also began to dominate regional faunal assemblages during the mid-Holocene, with a transition to a completely xerophilous fauna occurring by 2500 yr B.P. It is interesting to note that the transition from temperate to xeric faunas took such a long time to complete. Since there is good agreement between different sources of proxy data, one can only surmise that the climatic transition was indeed gradual in this region.

Recently I have begun research on late Quaternary insect faunas from the Colorado Plateau region of Utah and Arizona (Fig. 11.3, Nos. 10, 29, 47, 75, and 93) in collaboration with colleagues at Northern Arizona University. Preliminary results of samples ranging in age from 30,600 to 1500 yr B.P. suggest that late Wisconsin climatic conditions were cooler and moister than those at present and that the plateau supported a mosaic of grassland and shrub communities without modern analogue (Elias et al., 1992b).

Hall, Van Devender, and Olson (1988, 1989, 1990) have studied insect faunas from several sites in the Sonoran Desert of southwestern Arizona and northwestern Sonora, Mexico (Fig. 11.3, Nos. 2, 16, 43, 82, 102, 103, and 111). Unlike the records from the Chihuahuan Desert insect fauna, those from the Sonoran Desert fauna indicate little change in that all taxa recovered from Sonoran middens probably live within a few kilometers of the midden sites today. In their study of fossils from middens in Organ Pipe Cactus National Monument, Hall et al. (1990) noted that the fauna was only moderately diverse in the middle Holocene, despite increased precipitation during that interval. The record of the arthropod fauna showed a marked increase in diversity during the late Holocene, as more subtropical plants and warmer climates became established about 4000 yr B.P. The increase in diversity in fossil arthropod assemblages has been correlated with a decline in the frequency of winter freezes and secondarily with the amount of summer precipitation. Many warm-stenothermic insects probably dispersed into the study region from Sonora, Mexico, during the last 4000 yr. This late Holocene peak in species richness is in sharp contrast to the Chihuahuan Desert insect record, which showed the least number of species in late Holocene samples. The Chihuahuan Desert fauna may have been an important source of immigrant species into the Sonoran Desert in the late Holocene.

The fossil arthropod record from northwestern Sonora is similar to that of the Puerto Blanco Mountains of Organ Pipe Cactus National Monument in that it shows an increase in species richness in the late Holocene (Hall et al., 1988). A major difference is that the northwestern Sonoran record shows the establishment of a relatively modern fauna by the early Holocene. However, this conclusion is primarily based on identifications made only to the generic or family level. Additional specific identifications may help to clarify the issue of stable versus changing faunas in the Holocene. If the Sonoran Desert insect fauna truly was stable through the late Quaternary, then this represents a significant difference between the Sonoran and Chihuahuan Desert insect faunal histories. Since climatic factors appear to have been the most important element in bringing about the large-scale distributional changes seen in the Chihuahuan Desert faunas during the last 40,000 yr, this might suggest that the late Quaternary climatic regimes in the Sonoran Desert region were more stable than those of the Chihuahuan Desert.

Overview of North American Studies

Quaternary entomology has developed markedly in North America during the last twenty years. Paleoentomologists have learned a great deal about insect response to climatic change during the Wisconsin glaciation and the Holocene, especially in the central and northeastern United States, southeastern Canada, and the Chihuahuan Desert. However, much remains to be done. Extensive regions, including the prairies of central Canada, most of the Northwest Territories, the southern United States, the Great Basin, and the Pacific Coast states, are virtually unstudied. Also, very little is known about faunas predating the last interglacial.

Clearly, insects have responded to late Quaternary environmental changes in North America much as they have elsewhere, namely by making marked distributional shifts. As in Europe, certain suites of species appear again and again, having been repeatedly reassembled whenever or wherever suitable (cold or warm) climatic conditions prevailed. Some of these associations had great longevity during the Wisconsin interval and have only dissolved in the Holocene (Schwert, 1992). From a Quaternary perspective (i.e., the last 1.7 million years), our present biological communities are ephemeral associations of species, lasting only a small fraction of the life spans of the species involved. It seems that the environmental forces that shaped North American ecosystems during most of the last 100,000 years have held sway for most of the Quaternary. Cold environments have persisted through long glacial and interstadial episodes. Warm interglacials are short-lived. Moreover, the fossil record indicates that North American insects have responded rapidly and with great sensitivity to these shifting environmental patterns. Their almost constant geographic shifts may have prevented speciation in the Quaternary; extinction or speciation in the North American record of the last million years must have been very limited.

These are the same inferences that emerged from the study of the British fossil insect record. North American studies show that the British story was not exceptional or unique.

SOUTH AMERICAN STUDIES

> In recent years, the view that Pleistocene climatic events played a major role in the evolution of the biotas of southern, primarily tropical continents has begun to displace the previously held conviction that these areas were relatively stable during the Quaternary.
> —Beryl Vuilleumier (1971)

South America is the latest continent to receive the attention of Quaternary entomologists. Vuilleumier's seminal paper, although it did not discuss insect fossils (at that time, very little had been done) set the tone for much that has come afterward. Recent studies in South America have amply demonstrated that the biota of this continent underwent a series of dynamic changes in response to Pleistocene climates. In the tropical regions, the changes were mostly in the amount and seasonality of precipitation. In the higher latitudes of southern South America, changes in temperature and precipitation brought mountain glaciers and subantarctic conditions. Most Quaternary insect research in South America has been conducted in the latter region (Fig. 11.5 and Table 11.4).

Only two studies have dealt with fossil insect faunas from tropical latitudes in South America. Binford (1982) investigated the paleolimnology of Lake Valencia, in Venezuela (Fig. 11.5, No. 1), including chironomid larval remains in sediment cores. Basal sediments were radiocarbon-dated at about 12,500 yr B.P. The lake was relatively shallow and subject to drying-out until about 10,500 yr B.P., when it began to fill rapidly. Between 10,000 and 7000 yr B.P. sedimentation decreased, lake waters became less turbid, and chironomid species characteristic of deep, clear water were dominant.

From 7000 to 2200 yr B.P. the water level of the lake dropped below that of the outflowing streams; the lake increased in salinity, and most invertebrate populations declined. Lake levels and water quality have undergone a series of fluctuations in the last 1200 yr.

A second study in equatorial South America was an investigation of insect remains from tar seeps at Talara, Peru (Fig. 11.5, No. 7), by Churcher (1966). The mode of deposition of these tar seeps is similar to that of the asphalt deposits at Rancho La Brea, California, and the ecological composition of the insect faunas is likewise similar at the family level. The Talara faunas are

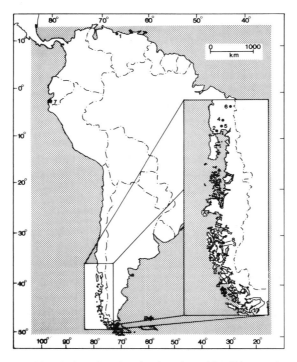

Figure 11.5. Map of South America showing location of fossil insect sites. Site numbers are keyed to the list of sites in Table 11.4.

Table 11.4

South American Quaternary insect fossil sites and references

Site	Reference(s)
1. Lake Valencia, Venezuela	Binford (1982)
2. Monte Verde, Chile	Ashworth et al. (1989)
3. Puerto Edén	Ashworth et al. (1991)
4. Puerto Octay	Hoganson and Ashworth (1992)
5. Puerto Varas	Hoganson and Ashworth (1992)
6. Rio Caunahue	Ashworth and Hoganson (1987)
7. Talara, Peru	Churcher (1966)
8. Témpano del Sur, Chile	Ashworth and Markgraf (1989)

dominated by aquatic beetles that lived in water just above the tar, scavengers that fed on carcasses of animals trapped in the tar, and terrestrial insects that strayed onto the tar surface and were trapped. The age range of the studied assemblages is between 14,400 and 13,600 yr B.P. Unfortunately, no specific determinations were made, in spite of remarkably good preservation, so Churcher was not able to make a paleoclimatic interpretation from the data.

Ashworth and Hoganson have studied a number of late glacial and Holocene insect faunas from the central and southern coastal regions of Chile (Fig. 11.5). As I discussed in Chapter 7, they examined insect fossils from essentially natural deposits at the Monte Verde archaeological site (Fig. 1.5, No. 2). At Monte Verde, insect faunas were extracted from a 13,000-yr B.P. peat horizon (Ashworth et al., 1989). The Monte Verde peats yielded abundant, diverse insect assemblages. The environmental reconstruction based on the insect fauna is a shallow creek with sparsely vegetated sand bars and some bogs, flowing through Valdivian (southern South American) rain forest. The late glacial paleoclimate was interpreted as being very similar to the modern climate; this interpretation is in agreement with all of the other lines of evidence except for the pollen interpretation, as discussed in Chapter 7.

To develop an adequate knowledge of the modern distribution and ecological requirements of the beetle fauna of central and southern Chile, Ashworth and Hoganson (1987) carried out an extensive collecting program for several field seasons, much of it in the Puyehue National Park, which contains a range of biological communities from lowland deciduous forest near sea level to alpine tundra above 1200 m elevation. Ashworth and Hoganson identified 462 species of beetles from 48 families in 41 locations in and around the park. This intensive study allowed them to develop meaningful interpretations of late Quaternary fossil assemblages.

Hoganson and Ashworth (1992) summarized their work on late Pleistocene insect faunas from south-central Chile. Based on fossil assemblages from sites at Puerto Octay (Fig. 11.5, No. 4), Puerto Varas (Fig. 11.5, No. 5), and Rio Caunahue (Fig. 11.5, No. 6), they postulated that cold-adapted beetles colonized regional lowlands following glacial retreat about 18,000 yr B.P. Depauperate moorland faunas dominated the interval 18,000–14,000 yr B.P.; these were replaced by more thermophilous taxa, including arboreal beetles. This amelioration continued until 12,500 yr B.P., by which time the beetle fauna was comprised solely of rain forest species.

This climatic amelioration was on the order of 4–5°C in mean annual temperature, and it brought a fivefold increase in beetle species diversity over the previous (glacial) fauna. This transition to postglacial conditions was the last major regional climatic change shown in the insect faunal sequence. As in the Monte Verde assemblages, Hoganson and Ashworth saw no evidence of a Younger Dryas cooling in this region.

To the south, Ashworth et al. (1991) and Ashworth and Markgraf (1989) studied late glacial and Holocene insect faunas at two sites in the Chilean Channel region. This region was heavily glaciated during the late Pleistocene, but these studies have shown that there must have been ice-free regions in which cold-adapted biota survived. At Puerto Edén (Fig. 11.5, No. 3), Ashworth described assemblages ranging in age from 13,000 yr B.P. to recent. Insects apparently invaded newly deglaciated terrain from regional refugia. The fossil data indicate that this colonization occurred within a few decades of deglaciation. The earliest beetle assemblages were composed of species associated with barren, open ground environments. The pioneering vegetation was a heath. However, both the paleobotanical and the insect data suggest that the pioneer heathland also supported shrubs and trees.

A southern beech woodland existed at Puerto Edén from 13,000 to 9500 yr B.P.; this supported a mixture of open ground and woodland beetles. This same environment was reflected by insect and paleobotanical assemblages dating from 11,200 to 10,100 yr B.P. from the nearby Témpano del Sur site (Fig. 11.5, No. 8) (Ashworth and Markgraf, 1989). No evidence was found in either the beetle fossils, pollen, or plant macrofossils for a proposed Younger Dryas cooling in the Chilean Channel region.

The paleoenvironmental sequence in the Holocene included a mosaic of rain forest and moorland habitats from 9500 to 5500 yr B.P., an increase in heathland importance from 5500 to 3000 yr B.P., and a return to a rain forest–moorland mosaic within the last 3000 yr. The mid-Holocene heathland episode has been attributed to increasing aridity, evidence for which has also been found in pollen spectra from Tierra del Fuego.

Ashworth and Hoganson have made a significant beginning in southern South America, overcoming tremendous obstacles just to develop an understanding of the modern fauna, let alone Quaternary assemblages. Their scientific curiosity led them to explore the history of life in a region that many would consider "the ends of the Earth." In order to identify and interpret fossil insect assemblages in this region, Ashworth and Hoganson spent years collecting modern specimens and tabulating their ecological requirements and the altitudinal zonation of their populations. This painstaking work paved the way for meaningful fossil studies.

Quaternary entomological research has made great strides in the Western Hemisphere during the past twenty years. Even so, perhaps the most striking aspect of research in North and South America is how much work remains to be done. Many regions remain completely unstudied. All studied regions would benefit from additional work, either to flesh out more data on different time intervals, to add clarity to reconstructions already made, or both. There is much left to be done, and most workers in the field find themselves inundated with samples and additional research opportunities. Time and money seem to be the only limiting factors.

12

CONCLUSIONS AND PROSPECTUS

> The importance of paleoentomological research has yet to be fully under-
> stood, and it still largely remains as one of the untapped areas of Quaternary
> research.
> —Alan and Anne Morgan (1987)

Quaternary entomology, like any fledgling field in science, has had its share of
detractors, doubters, and skeptics. Insect taxonomists have doubted that Quaternary fossils could be identified. Paleobotanists have doubted that fossil insect
data could be useful for paleoenvironmental reconstructions. Most of the criticisms
leveled at Quaternary entomology have been useful in that they have driven
investigators to sharpen their methods, scrutinize their data more thoroughly, and
thus make increasingly more convincing arguments. This heightened focus, in
turn, has led to a general acceptance of our research, which is no small achievement. Now paleobiologists and neobiologists alike are starting to ask us, "Given
that your method works, what more can it tell us?" Botanists are beginning to
inquire, "What can insect fossil assemblages tell us about the history of plant-insect
species interactions?" Ecologists wonder, "What mechanisms shape the dynamics
of ecosystems through time?" Climatologists are asking, "What can beetles tell us
about past moisture regimes and changes in seasonality?"

Now that the barrier of acceptance has been broken, these and other questions offer new challenges, and promise fascinating new collaborations with
colleagues from a host of other fields. With one foot planted in paleontology
and the other in modern biology, we are ready to bridge the gap, challenge the
old dogmas, and continue to learn from both sides of the time line. That is what
makes this type of interdisciplinary research so exciting.

Quaternary entomologists must continue to find and analyze assemblages of
Quaternary insects, with the particular goal of developing knowledge of unstudied regions and unstudied intervals of Quaternary time. However, some
new research approaches promise to take the field into the twenty-first century
with some excitement.

One of these approaches is the application of the mutual climatic range (MCR) method to faunal assemblages in regions outside Europe in order to establish an objective and repeatable quantitative assessment of Quaternary climatic changes. I am now beginning to apply this method to North American studies. As outlined in Chapter 11, a number of North American studies have focused on the Wisconsin-Holocene transition at sites near the late Wisconsin ice margins. Good regional syntheses of these studies have been published (Morgan and Morgan, 1980; Morgan et al., 1984a; Schwert and Ashworth, 1988; Schwert, 1992), but these efforts constitute mostly qualitative rather than quantitative paleoenvironmental reconstructions. The original investigations were carried out in six different laboratories by nine investigators, without a central, unifying research strategy. Quaternary entomology is by no means alone in this fragmentation, but the quantitative approach is more objective than previous methods. As such, the development of MCR analyses of North American data will probably give paleoentomology wider acceptance both within and outside the field of Quaternary science.

Another new approach in Quaternary entomology is the development of isotopic studies of fossil insect cuticle. Miller et al. (1988) published the results of studies on the hydrogen isotope composition of beetle cuticle. The hydrogen and carbon that are incorporated into the exoskeleton probably are derived ultimately (via the food chain and the insects' metabolism) from environmental carbon dioxide and water bound by photosynthesis into plant carbohydrates. This hypothesis is supported by carbon isotopic studies of laboratory-reared insects (Miller et al., 1985b). Hydrogen isotope (deuterium/hydrogen) results, obtained from chitin nitration studies, show small interspecific and intraspecific variation in beetles raised in different environments. Chitin nitrate samples from insects collected from across North America vary in a manner similar to the D/H ratios measured from plant cellulose nitrates when compared with environmental temperature, especially when above-freezing temperatures are considered.

These studies suggest that chitin may remain chemically stable enough to allow paleoecological and paleoclimatic studies based on stable isotope analysis (Miller, 1991). Further research may increase our understanding of the derivation of chitin isotopes and the ecological-environmental influences on their composition. Laboratory methods will need to improve, however, in order for isotopic analyses to be performed on the smaller amounts of chitin available in fossil assemblages.

An additional field that is currently gaining wide recognition is the extraction and amplification of fossil DNA (see Cherfas, 1991). I have begun a collaboration with Dr. Brian Farrell of Cornell University to extract fossil DNA from dried soft tissues extracted from beetle abdomens from Chihuahuan Desert packrat middens. Preservation of soft tissue is relatively rare in water-lain deposits, because most of the sclerites become disassociated and the soft parts decompose. However, many packrat middens preserve exoskeletons nearly intact, and, in the hot, dry air of the desert, soft tissues shrivel and dry before significant decomposition takes place. This study may provide the first tangible evidence on the longevity of beetle genotypes in the Quaternary. As with so many other aspects of Quaternary entomology, we can only begin to perceive where this field may lead.

Of all the strange and fascinating things to be learned from the study of Quaternary insects, perhaps none is as striking as what the insect remains have to tell us about insects themselves. To the untrained eye, they appear to be frail, fragile little creatures, barely worthy of attention unless they damage our crops or infest our houses. Nothing could be further from the truth. Insects may be small, but they are extremely successful. They are the ultimate survivors in the game of life that has been played out on this planet through hundreds of millions of years. The fossil record shows that insects essentially took the Pleistocene in stride, with the full complement of species emerging unscathed at the end of multiple glaciations. The same series of events wiped out whole groups of larger animals.

So Quaternary insect fossils are more than just convenient tracers of past environments. The fossil evidence speaks volumes about how genetics works in a large group of very mobile creatures. It shows that even though ecosystems appear to us to be stable entities, they are really very ephemeral on time scales of greater than a century or two. Meanwhile, the insects, the great survivors, keep moving, keep reproducing, keep succeeding. Who knows what else they could teach us, if we would pay attention?

Appendix

INSECTS AND OTHER ARTHROPODS MENTIONED IN THE TEXT

Scientific name	Common name
Class Insecta	Insects
Order Coleoptera	Beetles
Family	
Bostrichidae	Branch and twig boring beetles
Rhyzopertha dominica	
Buprestidae	Metallic wood-boring beetles
Byrrhidae	Pill beetles
Cytilus alternatus	
Morychus aeneus	
Morychus viridus	
Cantharidae	Soldier beetles
Carabidae	Ground beetles
Agonum bembidioides	
Agonum quadripunctatum	
Agonum quinquepunctatum	
Amara alpina	
Amara bokeri	
Amara chalcea	
Amara erratica	
Amara glacialis	
Asaphidion yukonense	
Bembidion dauricum	
Bembidion grisvardi	
Bembidion ibericum	
Bembidion rusticum	
Blethisa multipunctata	
Callistus lunatus	
Calosoma inquisitor	
Calosoma porosifrons	
Carabus chamissonis	

Scientific name	Common name
Carabus truncaticollis	
Chlaenius niger	
Diacheila arctica	
Diacheila polita	
Dyschirius frigidus	
Elaphrus americanus	
Elaphrus clairvillei	
Elaphrus parviceps	
Harpalus amputatus	
Harpalus fulvilabrus	
Nebria gyllenhalli	
Nebria nivalis	
Nebria trifaria	
Notiophilus borealis	
Oodes gracilis	
Patrobus septentrionis	
Patrobus stygicus	
Pelophila borealis	
Pelophila rudis	
Pterostichus brevicornis	
Pterostichus caribou	
Pterostichus haematopus	
Pterostichus majus	
Pterostichus nivalis	
Pterostichus pinguedineus	
Pterostichus sublaevis	
Pterostichus tundrae	
Pterostichus ventricosus	
Pterostichus vermiculosus	
Rhadine longicolle	
Trachypachus zetterstedti	
Trichocellus cognatus	
Trichocellus punctatellus	
Chrysomelidae	Leaf beetles
Chrysolina septentrionalis	
Chrysolina subsulcata	
Chrysomela tajmyrensis	
Plateumaris nitida	
Cicindelidae	Tiger beetles
Cicindela limbalis	
Cleridae	Checkered beetles

Scientific name	Common name
Coccinellidae	Ladybird beetles
Cryptophagidae	Silken fungus beetles
Atomaria cf. *apicalis*	
Curculionidae	Weevils or snout beetles
Lepidophorus lineaticollis	
Sitophilus granarius	Granary weevil
Sitophilus oryzae	Rice weevil
Stephanocleonus eruditus	
Trichalophus korotyaevi	
Vitavitus thulius	
Dermestidae	Dermestid beetles
Thylodrias contractus	Odd beetle
Dytiscidae	Predaceous diving beetles
Elateridae	Click beetles
Elmidae	Riffle beetles
Endomychidae	Handsome fungus beetles
Mycetaea hirta	
Heteroceridae	Variegated mud-loving beetles
Histeridae	Hister beetles
Hydrophilidae	Water scavenger beetles
Helophorus aquaticus	
Helophorus arcticus	
Helophorus lapponicus	
Helophorus mongoliensis	
Helophorus oblongus	
Helophorus obscurellus	
Helophorus sibiricus	
Helophorus splendidus	
Helophorus tuberculatus	
Hydroscaphidae	Skiff beetles
Lampyridae	Fireflies or lightningbugs
Lathridiidae	Minute brown scavenger beetles
Corticaria elongata	
Leiodidae	Round fungus beetles
Meloidae	Blister beetles
Lytta veriscatoria	Spanish fly
Mycetophagidae	Hairy fungus beetles
Typhaea stercorea	
Ptinidae	Spider beetles
Scarabaeidae	Dung beetles and chafers
Aphodius bonvouloiri	

Scientific name	Common name
Aphodius congregatus	
Aphodius holdereri	
Aphodius lapponum	
Canthon simplex	
Cetonia aurata	
Copris pristinus	
Onthophagus cochisus	
Onthophagus everestae	
Onthophagus lecontei	
Onthophagus massai	
Scolytidae	Bark and ambrosia beetles
Phloeotribus lecontei	Spruce bark beetle
Polygraphus rufipennis	
Scolytus koenigi	
Scolytus scolytus	Elm bark beetle
Silphidae	Carrion beetles
Nicrophorus marginatus	
Nicrophorus nigrita	
Silpha coloradensis	
Staphylinidae	Rove beetles
Anotylus gibbulus	
Atheta islandica	
Holoboreaphilus nordenskioeldi	
Micralymma brevilingue	
Micropeplus hoogendorni	
Micropeplus hopkinsi	
Omalium excavatum	
Tachinus apterus	
Tachinus arcticus	
Tachinus brevipennis	
Tachinus caelatus	
Tachinus frigidus	
Tachyporus canadensis	
Tachyporus rulomus	
Xylodromus concinnus	
Xylodromus depressus	
Tenebrionidae	Darkling beetles
Apsena laticornis	
Coniontis abdominalis	
Eleodes grandicollis	
Eleodes osculans	

Scientific name	Common name
Order Hymenoptera	Ants, wasps, and bees
Family	
Formicidae	Ants
Camponotus herculeanus	Carpenter ant
Formica	Red ant
Myrmecocystus	Honey ant
Order Hemiptera	Bugs
Family	
Corixidae	Water boatmen
Cydnidae	Burrower Bugs
Gerridae	Water striders
Lygaeidae	Seed bugs
Notonectidae	Backswimmers
Pentatomidae	Stink bugs
Reduviidae	Assassin bugs
Saldidae	Shore bugs
Order Homoptera	Hoppers, aphids, cicadas
Family	
Cicadelidae	Leaf hoppers
Order Diptera	Flies
Family	
Agromyzidae	Leafminer flies
Calliphoridae	Blowflies
Drosophilidae	Fruitflies
Drosophila	
Muscidae	House flies, stable flies
Fannia scalaris	Latrine fly
Chaoboridae	Phantom midges
Chironomidae	Midges
Oestridae	Bot flies, warble flies
Cobboldia russanovi	
Sciaridae	Dark-winged fungus gnats
Tipulidae	Craneflies
Order Isoptera	Termites
Family	
Rhinotermitidae	
Reticulitermes cf. *tibialis*	
Order Anoplura	Sucking lice
Pediculidae	Head and body lice
Pediculus humanis capitis	Human head louse

Scientific name	Common name
Pediculus humanus humanus	Human body louse
Phthiridae	Crab lice
Phthirus pubis	Crab louse
Class Arachnida	Spiders, mites, ticks, centipedes, scorpions
Order Araneae	Spiders
Family	
Linyphiidae	Sheet-web spiders
Order Acari	Mites and ticks
Family	
Trombiculidae	Harvest mites, chiggers
Trombicula	Chigger
Lardoglyphidae	Stored product mites
Lardoglyphus robustisetosus	
Suborder Oribatida	Oribatid mites

REFERENCES

Aalto, M. M., Coope, G. R., and Gibbard, P. L. 1984. Late Devensian river deposits beneath the Floodplain Terrace of the River Thames at Abingdon, Berkshire, England. *Proceedings of the Geologists' Association* 95:65–79.

Addyman, P. V. 1989. The archaeology of public health at York, England. *World Archaeology* 21:244–264.

Addyman, P. V., Hood, J. S. R., Kenward, H. K., MacGregor, A., and Williams, D. 1976. Palaeoclimate in urban environmental archaeology at York, England: Problems and potential. *World Archaeology* 8:220–233.

Alfieri, A. 1931. Les insectes de la tombe de Toutankhamon. *Bulletin de la Société Royale Entomologique d'Egypte* 3/4:188–189.

Ami, H. M. 1894. Fossil insects from the Leda Clays of Ottawa and vicinity. *The Ottawa Naturalist* 9:190–191.

Ammann, B., Chaix, L., Eicher, U., Elias, S. A., Gaillard, M.-J., Hofmann, W., Siegenthaler, U., Tobolski, K., and Wilkinson, B. 1983. Vegetation, insects, mollusks and stable isotopes in Late-Würm deposits at Lobsigensee (Swiss Plateau). *Revue de Paléobiologie* 2:184–204.

Anderson, E. 1984. Who's who in the Pleistocene: A mammalian beastiary. In Martin, P. S., and Klein, R. G. (eds.), *Quaternary Extinctions: A Prehistoric Revolution.* Tucson: University of Arizona Press, pp. 40–89.

Anderson, R. S., and Peck, S. B. 1985. *The Insects and Arachnids of Canada,* Part 13: *The Carrion Beetles of Canada and Alaska (Coleoptera: Silphidae and Agyrtidae).* Agriculture Canada, Research Branch Publication 1778. 121 pp.

Anderson, R. S., Miller, N. G., Davis, R. B., and Nelson, R. E. 1990b. Terrestrial fossils in the marine Presumpscot Formation: Implications for Late Wisconsinan paleoenvironments and isostatic rebound along the coast of Maine. *Canadian Journal of Earth Sciences* 27:1241–1246.

Anderson, T. W., Matthews, J. V., Jr., Mott, R. J., and Richard, S. H. 1990a. The Sangamonian Pointe–Fortune site, Ontario-Quebec border. *Géographie physique et Quaternaire* 44:271–287.

Andersson, G. 1898. Studier öfver Finlands törfmossar och fossila kvartärflora. *Fennia* 15:3–30.

Angus, R. B. 1973. Pleistocene *Helophorus* (Coleoptera: Hydrophilidae) from Borislav and Starunia in the western Ukraine, with a reinterpretation of M. Lomnicki's species, description of a new Siberian species, and comparison with British Weichselian faunas. *Philosophical Transactions of the Royal Society of London, Series B* 265:299–326.

Angus, R. B. 1975. Fossil Coleoptera from Weichselian deposits at Voorthuizen, The Netherlands. *Geologie en Mijnbouw* 54:146–147.

Angus, R. B. 1983. Evolutionary stability since the Pleistocene illustrated by reproductive compatibility between Swedish and Spanish *Helophorus lapponicus* Thompson (Coleoptera: Hydrophilidae). *Biological Journal of the Linnean Society of London* 19:17–25.

Arkhipov, S. A., Isayeva, L. L., Bespaly, V. G., and Glushkova, O. 1986. Glaciation of Siberia and north-east USSR. *Quaternary Science Reviews* 5:463–474.

Armstrong, D. M. 1982. *Mammals of the Canyon Country: A Handbook of Mammals of Canyonlands National Park and Vicinity.* Moab, Utah: Canyonlands Natural History Association. 86 pp.

Ashworth, A. C. 1972. A Late-Glacial insect fauna from Red Moss, Lancashire, England. *Entomologica Scandinavica* 3:211–224.

Ashworth, A. C. 1973. The climatic significance of a Late Quaternary fauna from Rodbaston Hall, Staffordshire, England. *Entomologica Scandinavica* 4:191–205.

Ashworth, A. C. 1977. A late Wisconsin Coleopterous assemblage from southern Ontario and its environmental significance. *Canadian Journal of Earth Sciences* 14:1625–1634.

Ashworth, A. C. 1979. Quaternary Coleoptera studies in North America: Past and present. In Erwin, T. L., Ball, G. E., and Whitehead, D. R. (eds.), *Carabid Beetles: Their Evolution, Natural History, and Classification.* The Hague: Dr. W. Junk, pp. 395–406.

Ashworth, A. C. 1980. Environmental implications of a beetle assemblage from the Gervais Formation (Early Wisconsinan?), Minnesota. *Quaternary Research* 13:200–212.

Ashworth, A. C., and Brophy, J. A. 1972. Late Quaternary fossil beetle assemblage from the Missouri Coteau, North Dakota. *Geological Society of America Bulletin* 83:2981–2988.

Ashworth, A. C., and Hoganson, J. W. 1987. Coleoptera bioassociations along an elevational gradient in the lake region of southern Chile, and comments on the postglacial development of the fauna. *Annals of the Entomological Society of America* 80:865–895.

Ashworth, A. C., and Markgraf, V. 1989. Climate of the Chilean Channels between 11,000 and 10,000 yr B.P. based on fossil beetle and pollen analyses. *Revista Chilena de Historia Natural* 62:61–74.

Ashworth, A. C., Clayton, L., and Bickley, W. B. 1972. The Mosbeck site: A paleoenvironmental interpretation of the Late Quaternary history of Lake Agassiz based on fossil insect and mollusk remains. *Quaternary Research* 2:176–186.

Ashworth, A. C., Schwert, D. P., Watts, W. A., and Wright, H. E., Jr. 1981. Plant and insect fossils at Norwood in south-central Minnesota: A record of late-glacial succession. *Quaternary Research* 16:66–79.

Ashworth, A. C., Hoganson, J., and Gunderson, M. 1989. Fossil beetle analysis. In Dillehay, T. D. (ed.), *Monte Verde: A Late Pleistocene Settlement in Chile,* Volume 1: *Palaeoenvironment and Context.* Washington, D.C.: Smithsonian Institution Press, pp. 211–226.

Ashworth, A. C., Markgraf, V., and Villagran, C. 1991. Late Quaternary climatic history of the Chilean Channels based on fossil pollen and beetle analyses, with an analysis of the modern vegetation and pollen rain. *Journal of Quaternary Science* 6:279–291.

Askevold, I. S. 1990. Classification of Tertiary fossil Donaciinae of North America and their implications about evolution of Donaciinae (Coleoptera: Chrysomelidae). *Canadian Journal of Zoology* 68:2135–2145.

Atkinson, T. C., Briffa, K. R., Coope, G. R., Joachim, M., and Perry, D. W. 1986. Climatic calibration of coleopteran data. In Berglund, B. (ed.), *Handbook of Holocene Palaeoecology and Palaeohydrology.* New York: Wiley, pp. 851–858.

Atkinson, T. C., Briffa, K. R., and Coope, G. R. 1987. Seasonal temperatures in Britain during the last 22,000 years, reconstructed using beetle remains. *Nature* 325:587–592.

Baker, A. S. 1990. Two new species of *Lardoglyphus* Oudemans (Acari: Lardoglyphidae) found in the gut contents of human mummies. *Journal of Stored Products Research* 26:139–147.

Baker, R. G., Rhodes, S. R., Schwert, D. P., Ashworth, A. C., Frest, T. J., Hallberg, G. R., and Janssens, J. A. 1986. A full-glacial biota from southeastern Iowa, USA. *Journal of Quaternary Science* 1:91–108.

Baker, R. G., Schwert, D. P., Bettis, E. A., III, Kemmis, T. J., Horton, D. G., and Semken, H. A. 1991. Mid-Wisconsinan stratigraphy and paleoenvironments at the St. Charles site in south-central Iowa. *Geological Society of America Bulletin* 103:210–220.

Ball, G. E. 1966. A revision of the North American species of the subgenus *Cryobius* Chaudoir (*Pterostichus,* Carabidae, Coleoptera). *Opuscula Entomologica Supplement* 28. 166 pp.

Barnowsky, A., Barnowsky, C. W., Nickmann, R. J., Ashworth, A. C., Schwert, D. P., and Lantz, S. W. 1988. Late Quaternary paleoecology at the Newton site, Bradford Co., northeastern Pennsylvania: *Mammuthus columbi,* palynology, and fossil insects. *Bulletin of the Buffalo Society of Natural Sciences* 33:173–184.

Barr, T. C. 1960. The cavernicolous beetles of the subgenus *Rhadine,* genus *Agonum* (Coleoptera: Carabidae). *American Midland Naturalist* 64:45–65.

Bartlein, P. J., and Prentice, I. C. 1989. Orbital variations, climate and paleoecology. *Trends in Ecology and Evolution* 4:195–199.

Barry, R. G. 1982. Paleoclimate. In Hopkins, D. M., Matthews, J. V., Jr., Schweger, C. E., and Young, S. B. (eds.), *Paleoecology of Beringia.* New York: Academic Press, pp. 193–204.

Basilewsky, P. 1970. Les Coléoptères Carabidae de l'Ile d'Aldabra (Océan Indien). *Bulletin Annuel de la Société Royale Entomologique de Belgique* 106:221–222.

Bell, F. G., Coope, G. R., Rice, R., and Riley, T. 1972. Mid-Weichselian fossil-bearing deposits at Syston, Leicestershire. *Proceedings of the Geological Association* 83:197–211.

Bennet, K. D. 1990. Milankovitch cycles and their effects on species in ecological and evolutionary time. *Paleobiology* 16:11–21.

Berglund, B. E., and Digerfeldt, G. 1970. A paleoecological study of the Late-Glacial lake at Torreberga, Scania, South Sweden. *Oikos* 21:98–128.

Berglund, B. E., Lemdahl, G., Leidberg-Jonsson, B., and Persson, T. 1984. Biotic response to climatic changes during the time span 13,000–10,000 BP—A case from SW Sweden. In Mörner, N. A., and Karlén, W. (eds.), *Climatic Changes on a Yearly to Millenial Basis*. Dordrecht, Netherlands: Reidel, pp. 25–36.

Berman, D. I. 1990. Ecology of *Morychus viridis* (Coleoptera, Byrrhidae), a moss beetle from Pleistocene deposits in the northeastern USSR. In Kotlyakov, V. M., and Sokolov, V. E. (eds.), *Arctic Research: Advances and Prospects. Proceedings of the Conference of Arctic and Nordic Countries on Coordination of Research in the Arctic*. Moscow: Nauka, pp. 281–288.

Bidashko, F. G., and Proskurin, K. P. 1988. The entomological and carpological reconstruction of the bio-environment of the Singilian (Middle Pleistocene) of the lower Volga. *Paleontological Journal* 4:66–72.

Binford, M. W. 1982. Ecological history of Lake Valencia, Venezuela: Interpretation of animal microfossils and some chemical, physical, and geological features. *Ecological Monographs* 52:307–333.

Bishop, W. W., and Coope, G. R. 1977. Stratigraphical and faunal evidence for Lateglacial and Early Flandrian environments in south-west Scotland. In Gray, J. M., and J. J. Lowe (eds.), *Studies in the Scottish Lateglacial Environment*. Oxford: Pergamon Press, pp. 61–88.

Blake, W., Jr., and Matthews, J. V., Jr. 1979. New data on an interglacial peat deposit near Makinson Inlet, Ellesmere Island, District of Franklin. *Geological Survey of Canada, Papers* 79–1A:157–164.

Böcher, J. 1988. Insektundersøgelserne. In Grønnow, B., and Meldgaard, M. (eds.), Boplads i dybfrost [Settlements in permafrost]. *Naturens Verden* 11/12:409–423.

Böcher, J. 1989a. Boreal insect in northernmost Greenland: Palaeoentomological evidence from the Kab København Formation (Plio-Pleistocene), Peary Land. *Fauna norvegica Series B* 36:37–43.

Böcher, J. 1989b. First record of an interstadial insect from Greenland: *Amara alpina* (Paykull, 1790) (Coleoptera: Carabidae). *Boreas* 18:1–4.

Böcher, J. 1990. A two-million-year-old insect fauna from north Greenland indicating boreal conditions at the Plio-Pleistocene boundary. *Proceedings of the International Conference on the Role of Polar Regions in Global Change* 2:582–584.

Böcher, J., and Bennike, O. 1991. Interglacial land biotas of Jameson Land, East Greenland. *LUNQUA Report* 33:129–133.

Bohncke, S., Vandenberghe, J., Coope, G. R., and Reiling, R. 1987. Geomorphology and palaeoecology of the Mark valley (southern Netherlands): Palaeoecology, palaeohydrology and climate during the Weichselian Late Glacial. *Boreas* 16:69–85.

Bolton, J. 1862. On a deposit with insects, leaves, etc. near Ulverston. *Proceedings of the Geological Society of London* 18:274–277.

Borror, D. J., and White, R. E. 1970. *A Field Guide to the Insects.* Boston: Houghton Mifflin. 404 pp.

Borror, D. J., DeLong, D. M., and Triplehorn, C. A. 1981. *An Introduction to the Study of Insects,* Fifth Edition. Philadelphia: Saunders. 928 pp.

Bowen, D. Q., Rose, J., McCabe, A. M., and Sutherland, D. G. 1986. Correlation of Quaternary glaciations in England, Ireland, Scotland and Wales. *Quaternary Science Reviews* 5:299–340.

Bresciani, J., Haarlov, N., Nansen, P., and Moller, G. 1983. Head louse (*Pediculus humanus* subsp. *capitis* de Geer) from mummified corpses of Greenlanders, A.D. 1460 (+50). *Acta Entomologica Fennica* 42:24–27.

Bresciani, J., Haarlov, N., Nansen, P., and Moller, G. 1989. Head lice in mummified Greenlanders from A.D. 1475. In Hart Hensen, J. P., and Gullov, H. C. (eds.), *The Mummies from Qilakitsoq—Eskimos in the 15th Century. Meddelelser om Grønland, Man & Society* 12:89–92.

Briggs, D. J., Coope, G. R., and Gilbertson, D. D. 1975a. Late Pleistocene terrace deposits at Beckford, Worcestershire, England. *Geological Journal* 10:1–16.

Briggs, D. J., Gilbertson, D. D., Goudie, A. S., Osborne, P. J., Osmaston, H. A., Pettit, M. M. E., Shotton, F. W., and Stuart, A. J. 1975b. New interglacial site at Sugworth. *Nature* 257:477–479.

Briggs, D. J., Coope, G. R., and Gilbertson, D. D. 1985. The chronology and environmental framework of Early Man in the Upper Thames Valley. *British Archaeological Reports* 137. 176 pp.

Brubaker, L. B. 1986. Responses of tree populations to climatic change. *Vegetatio* 68:119–130.

Bryson, R. A. 1966. Air masses, streamlines and the boreal forest. *Geographical Bulletin* 8:228–269.

Buckland, P. C. 1973. Archaeology and environment in the Vale of York. *South Yorkshire Studies in Archaeology and Natural History* (Doncaster Museum) 1:6–18.

Buckland, P. C. 1974. Archaeology and environment in York. *Journal of Archaeological Science* 1:303–316.

Buckland, P. C. 1976a. The use of insect remains in the interpretation of archaeological environments. In Davidson, D. A., and Shackley, M. L. (eds.), *Geoarchaeology: Earth Science and the Past.* Boulder, Colorado: Westview Press, pp. 360–396.

Buckland, P. C. 1976b. *Niptus hololeucus* Fald. (Col., Ptinidae) from Roman deposits in York. *Entomologist's Monthly Magazine* 111:233–234.

Buckland, P. C. 1979. Thorne Moors: A palaeoecological study of a Bronze Age site (a contribution to the history of the British insect fauna). University of Birmingham, Department of Geography Occasional Publication 8. 173 pp.

Buckland, P. C. 1981a. The early dispersal of insect pests of stored products as indicated by archaeological records. *Journal of Stored Product Research* 17:1–12.

Buckland P. C. 1981b. An insect fauna from a Roman well at Empingham, Rutland. *Transactions of the Leicestershire Archaeological and Historical Society* 60:1–6

Buckland, P. C. 1984. North-west Lincolnshire 10,000 years ago. In Field, N., and Norwich, A. (eds.), *A Prospect of Lincolnshire*. Lincoln, pp. 11–17.

Buckland, P. C., and Coope, G. R. 1991. *A Bibliography and Literature Review of Quaternary Entomology*. Sheffield, England: Collins. 85 pp.

Buckland, P. C., and Kenward, H. K. 1973. Thorne Moor: A palaeoecological study of a Bronze Age site. *Nature* 241:405–406.

Buckland, P. C., Greig, J. R. A., and Kenward, H. K. 1974. York: An early medieval site. *Antiquity* 48:25–33.

Buckland, P. C., Holdsworth, P., and Monk, M. 1976. The interpretation of a group of Saxon pits in Southampton. *Journal of Archaeological Science* 3:61–69.

Buckland, P. C., Gerrard, A. J., Larsen, G., Perry, D. W., Savory, D. R., and Sveinbjarnardóttir, G. 1986a. Late Holocene palaeoecology at Ketilsstadir in Myrdalur, South Iceland. *Jökull* 36:41–55.

Buckland, P. C., Perry, D. W., Gislason, G. M., and Dugmore, A. J. 1986b. The pre-Landnám Fauna of Iceland: A palaeontological contribution. *Boreas* 15:173–184.

Buckland, P. C., Beal, C. J., and Heal, S. V. E. 1990. Recent work on the archaeological and environmental context of the Ferriby boats. In Ellis, S., and Crowther, D. R. (eds.), *Humber Perspectives: A Region through the Ages*. Hull: Hull University Press, pp. 131–146.

Buckland, P. C., Dugmore, A. J., Perry, D. W., Savory, D., and Sveinbjarnardóttir, G. 1991. Holt in Eyjafjallasveit, Iceland. A paleoecological study of the impact of Landnám. *Acta Archaeologica* 61:252–271.

Buckland, P. C., Sveinbjarnardóttir, G., Savory, D., McGovern, T. H., Skidmore, P., and Andreasen, C. 1983. Norsemen at Nipáitsoq, Greenland: A palaeoecological investigation. *Norwegian Archaeological Review* 16:86–98.

Callen, E. O. 1967. Analysis of the Tehuacan coprolites. In Byers, D. S. (ed.), *The Prehistory of the Tehuacan Valley,* Volume One: *Environment and Subsistence.* Austin: University of Texas Press, pp. 261–289.

Callen, E. O. 1970. Diet as revealed by coprolites. In Brothwell, D., and Higgs, E. (eds.), *Science in Archaeology*. New York: Praeger, pp. 235–243.

Cameron, A. W. 1958. Mammals of the islands in the Gulf of St. Lawrence. *National Museums of Canada Bulletin* 154:1–165.

Campbell, J. M. 1979. A revision of the genus *Tachyporus* Gravenhorst (Coleoptera: Staphylinidae) of North and Central America. Memoirs of the Entomological Society of Canada 109. 95 pp.

Campbell, J. M. 1980. Distribution patterns of Coleoptera in eastern Canada. *Canadian Entomologist* 112:1161–1175.

Campbell, J. M. 1982. A revision of the North American Omaliinae (Coleoptera: Staphylinidae). 3. The genus *Acidota* Stephens. *Canadian Entomologist* 114:1003–1029.

Campbell, J. M. 1983. A revision of the North American Omaliinae (Coleoptera: Staphylinidae): The genus *Olophrum* Erichson. *Canadian Entomologist* 115:577–622.

Campbell, J. M. 1988. New species and records of North American *Tachinus* Gravenhorst (Coleoptera: Staphylinidae). *Canadian Entomologist* 120:231–295.

Carter, K. D. 1985. Middle and Late Wisconsinan (Pleistocene) Insect Assemblages from Illinois. Masters thesis, Department of Geology, University of North Dakota, Grand Forks.

Cherfas, J. 1991. Ancient DNA: Still busy after death. *Science* 253:1354–1356.

Chomko, S. A., and Gilbert, B. M. 1991. Bone refuse and insect remains: Their potential for temporal resolution of the archaeological record. *American Antiquity* 56:680–686.

Chowne, P., Girling, M., and Greig, J. 1986. Excavations at an Iron Age defended enclosure at Tattershall Thorpe, Lincolnshire. *Proceedings of the Prehistoric Society* 52:159–188.

Chu, H. F., and Wang, L.-Y. 1975. Insect carcasses unearthed from Chinese antique tombs. *Acta Entomologica Sinica* 18:333–337 (in Chinese).

Churcher, C. S. 1966. The insect fauna from the Talara tar-seeps, Peru. *Canadian Journal of Zoology* 44:985–993.

CLIMAP members. 1981. Seasonal reconstructions of the earth's surface at the last glacial maximum. *Geological Society of America Map and Chart Series* MC-36:1–18.

Cockerell, T. D. A. 1911. Obituary of Samuel Hubbard Scudder. *Science* 34:338–342.

COHMAP members. 1988. Climatic changes of the last 18,000 years: Observations and model simulations. *Science* 241:1043–1052.

Cole, K. L. 1985. Past rates of change, species richness and a model of vegetational inertia in the Grand Canyon, Arizona. *American Naturalist* 125:289–303.

Coles, J. M. 1987. Tracks across the Wetlands: Multi-disciplinary studies of the Somerset Levels of England. In Coles, J. M., and Lawson, A. J. (eds.), *European Wetlands in Prehistory.* Oxford: Clarendon Press, pp. 145–167.

Colinvaux, P. 1987. Amazon diversity in light of the paleoecological record. *Quaternary Science Reviews* 6:93–114.

Colledge, S. M., and Osborne, P. J. 1980. Plant and insect remains. In Carver, M. (ed.), The excavation of three medieval craftsmen's tenements in Sidbury, Worcester. *Transactions of the Worcestershire Archaeological Society* 7:207–210.

Coope, G. R. 1959. A Late Pleistocene insect fauna from Chelford, Cheshire. *Proceedings of the Royal Society of London, Series B* 151:70–86.

Coope, G. R. 1961. On the study of glacial and interglacial insect faunas. *Proceedings of the Linnean Society of London* 172:62–71.

Coope, G. R. 1962a. Coleoptera from a peat interbedded between two boulder clays at Burnhead near Airdrie. *Transactions of the Geological Society of Glasgow* 24:279–286.

Coope, G. R. 1962b. A Pleistocene coleopterous fauna with arctic affinities from Fladbury, Worcestershire. *Geological Society of London, Quarterly Journal* 118:103–123.

Coope, G. R. 1965. Fossil insect faunas from Late Quaternary deposits in Britain. *Advancement of Science, 1965.* 564–575.

Coope, G. R. 1968a. Coleoptera from the "Arctic Bed" at Barnwell Station, Cambridge. *Geological Magazine* 105:482–486.

Coope, G. R. 1968b. An insect fauna from Mid-Weichselian deposits at Brandon, Warwickshire. *Philosophical Transactions of the Royal Society of London, Series B* 254:425–456.

Coope, G. R. 1968c. Fossil beetles collected by James Bennie from Late Glacial silts at Corstorphine, Edinburgh. *Scottish Journal of Geology* 4:339–348.

Coope, G. R. 1968d. Insect remains from silts below till at Garfield Heights, Ohio. *Geological Society of America Bulletin* 79:753–756.

Coope, G. R. 1969. Insect remains from Mid-Weichselian deposits at Peelo, the Netherlands. *Mededelingen Rijks Geologische Dienst NS* 20:79–83.

Coope, G. R. 1970. Interpretations of Quaternary insect fossils. *Annual Review of Entomology* 15:97–120.

Coope, G. R. 1971a. The fossil coleoptera from Glen Ballyre and their bearing upon the interpretation of Late Glacial environments. In Thomas, G. P. S. (ed.), *Field Guide to the Isle of Man*. Liverpool, England: Quaternary Research Association, pp. 13–15.

Coope, G. R. 1971b. Insecta. In Colhoun, E. A., and Mitchell, G. F. (eds.), *Interglacial Marine Formation and Lateglacial Freshwater Formation in Shortalstown Townland, Co. Wexford. Proceedings of the Royal Irish Academy* 71:234–238.

Coope, G. R. 1973. Tibetan species of dung beetle from Late Pleistocene deposits in England. *Nature* 245:335–336.

Coope, G. R. 1974a. Interglacial Coleoptera from Bobbitshole, Ipswich, Suffolk. *Journal of the Geological Society* 130:333–340.

Coope, G. R. 1974b. Report on the Coleoptera from Wretton. In West, R. G., Dickson, C. A., Catt, J. A., Weir, A. H., and Sparks, B. W., Late Pleistocene deposits at Wretton, Norfolk. II. Devensian deposits. *Philosophical Transactions of the Royal Society of London, Series B* 267:414–418.

Coope, G. R. 1975. Climatic fluctuations in northwest Europe since the Last Interglacial, indicated by fossil assemblages of Coleoptera. *Geological Journal Special Issue* 6:153–168.

Coope, G. R. 1976. Assemblages of fossil Coleoptera from terraces of the Upper Thames near Oxford. In Roe, D. *Quaternary Research Association Field Guide to the Oxford Region*. Oxford: Quaternary Research Association, pp. 20–23.

Coope, G. R. 1977. Fossil Coleopteran assemblages as sensitive indicators of climatic changes during the Devensian (Last) cold stage. *Philosophical Transactions of the Royal Society of London, Series B* 280:313–340.

Coope, G. R. 1978. Constancy of insect species versus inconstancy of Quaternary environments. In Mound, L. A., and Waloff, N. (eds.), *Diversity of Insect Faunas. Royal Entomological Society of London Symposium* 9:176–187.

Coope, G. R. 1979. Late Cenozoic fossil Coleoptera: Evolution, biogeography and ecology. *Annual Review of Ecology and Systematics* 10:247–267.

Coope, G. R. 1980. Appendix 1. Coleoptera from Late Glacial and Early Flandrian deposits at Folkestone. *Philosophical Transactions of the Royal Society of London, Series B* 291:38–39.

Coope, G. R. 1981. Report on the coleoptera from an eleventh-century house at Christ Church Place, Dublin. In Bekker-Nielson, H., Foote, P., and Olsen, O. (eds.),

Proceedings of the Eighth Viking Congress (1977). Odense, Denmark: Odense University Press, pp. 51–56.

Coope, G. R. 1982. Coleoptera from two Late Devensian sites in the lower Colne Valley, West London, England. *Quaternary Newsletter* 38:1–6.

Coope, G. R. 1986. Coleoptera analysis. In Berglund, B. E. (ed.), *Handbook of Holocene Palaeoecology and Palaeohydrology*. New York: Wiley, pp. 703–713.

Coope, G. R. 1987a. Fossil beetle assemblages as evidence for sudden and intense climatic changes in the British Isles during the last 45,000 years. In Berger, W. H., and Labeyrie, L. D. (eds.), *Abrupt Climatic Change*. Dordrecht, Netherlands: Reidel, pp. 147–150.

Coope, G. R. 1987b. The response of late Quaternary insect communities to sudden climatic changes. In Gee, J. H. R., and Giller, P. S. (eds.), *Organization of Communities*. 27th Symposium of the British Ecological Society. Aberystwyth, Wales: British Ecological Society, pp. 421–438.

Coope, G. R. 1988. Lateglacial climates in northwestern Europe interpreted from beetle assemblages. In *Tenth Biennial American Quaternary Association Meeting, Amherst, Massachusetts, Program and Abstracts*, p. 16.

Coope, G. R. 1990. The invasion of northern Europe during the Pleistocene by Mediterranean species of Coleoptera. In Di Castri, F., Hansen, A. J., and Debussche, M. (eds.), *Biological Invasions in Europe and the Mediterranean Basin*. Dordrecht, Netherlands: Kluwer, pp. 203–215.

Coope, G. R., and Angus, R. B. 1975. An ecological study of a temperate interlude in the middle of the Last glaciation, based on fossil Coleoptera from Isleworth, Middlesex. *Journal of Animal Ecology* 44:365–391.

Coope, G. R., and Brophy, J. 1972. Late Glacial environmental changes indicated by a coleopteran succession from North Wales. *Boreas* 1:97–143.

Coope, G. R., and Joachim, M. J. 1980. Lateglacial environmental changes interpreted from fossil Coleoptera from St. Bees, Cumbria, NW England. In Lowe, J. J., Gray, J. M., and Robinson, J. E. (eds.), *Studies in the Lateglacial of Northwest Europe*. Oxford: Pergamon Press, pp. 55–68.

Coope, G. R., and Osborne, P. J. 1968. Report on the coleopterous fauna of the Roman well at Barnsley Park, Gloucestershire. *Transactions of the Bristol and Gloucestershire Archaeological Society* 86:84–87.

Coope, G. R., and Sands, C. H. 1966. Insect faunas of the last glaciation from the Tame Valley, Warwickshire. *Proceedings of the Royal Society of London, Series B* 165:389–412.

Coope, G. R., and Tallon, P. W. J. 1983. A full glacial insect fauna from the Lea Valley, Enfield, North London. *Quaternary Newsletter* 40:7–12.

Coope, G. R., Shotton, F. W., and Strachan, I. 1961. A Late Pleistocene fauna and flora from Upton Warren, Worcestershire. *Philosophical Transactions of the Royal Society of London, Series B* 244:379–417.

Coope, G. R., Dickson, J. H., McCutcheon, J. A., and Mitchell, G. A. 1979. The Lateglacial and early Postglacial deposit at Drumurcher, Co. Monaghan. *Proceedings of the Royal Irish Academy, Series B* 79:63–85.

Coope, G. R., Jones, R. L., Keen, D. H., and Waton, P. V. 1985. The flora and fauna of Late Pleistocene deposits in St. Aubins Bay, Jersey, Channel Islands. *Proceedings of the Geologists' Association* 96:315–324.

Coope, G. R., Dickson, J. H., Jones, R. L., and Keen, D. H. 1987. The flora and fauna of late Pleistocene deposits on the Cotentin Peninsula, Normandy. *Philosophical Transactions of the Royal Society of London, Series B* 315:231–265.

Crossley, R. 1984. Fossil termite mounds associated with stone artifacts in Malawi, Central Africa. *Palaeoecology of Africa and of the Surrounding Islands and Antarctica* 16:397–401.

Crowson, R. A. 1981. *The Biology of the Coleoptera*. New York: Academic Press. 802 pp.

Curry, A. 1979. The insects associated with the Manchester mummies. In David, R. A. (ed.), *The Manchester Mummy Project*. Manchester, England: Manchester University Press, pp. 113–118.

Damblon, F., Coope, G. R., and Osborne, P. J. 1977. Paleoecological studies of peat bogs in the High Ardenne. In *Proceedings of the X INQUA Congress, Birmingham*. Norwich, England: Geobooks, p. 101.

Danks, H. V. 1978. Modes of seasonal adaptation in the insects. 1. Winter survival. *Canadian Entomologist* 110:1167–1205.

Danks, H. V. 1979. Terrestrial habitats and distributions of Canadian insects. In Danks, H. V. (ed.), *Canada and Its Insect Fauna. Memoirs of the Entomological Society of Canada* 108:195–210.

Danks, H. V. 1981. *Arctic Arthropods*. Ottawa: Entomological Society of Canada. 592 pp.

Dansgaard, W. 1987. Ice core evidence of abrupt climatic changes. In Berger, W. H., and Labeyrie, L. D. (eds.), *Abrupt Climatic Change*. Dordrecht, Netherlands: Reidel, pp. 223–233.

Dansgaard, W., White, J. W. C., and Johnson, S. J. 1989. The abrupt termination of the Younger Dryas climate event. *Nature* 339:532–534.

Darlington, P. J. 1938. The American Patrobini. *Entomologica Americana* 18:135–183.

David, A. R. 1984. The background. In David, A. R., and Topp, E. (eds.), *Evidence Embalmed: Modern Medicine and the Mummies of Ancient Egypt*. Manchester, England: Manchester University Press, pp. 35–36.

Davis, M. B., Woods, K. D., Webb, S. L., and Futyma, R. P. 1986. Dispersal versus climate: Expansion of *Fagus* and *Tsuga* into the Upper Great Lakes region. *Vegetatio* 67:93–103.

de Beaulieu, J. L., Pons, A., and Reille, M. 1982. Recherches pollenanalytiques sur l'histoire de la végétation de la bordure nord du massif du Cantal (Massif Central, France). *Pollen et Spores* 34:251–300.

de Beaulieu, J. L., Clerc, J., and Reille, M. 1984. Late Weichselian fluctuations in the French Alps and Massif Central from pollen analysis. In Mörner, N. A., and Karlén, W. (eds.), *Climatic Changes on a Yearly to Millenial Scale*. Dordrecht, Netherlands: Reidel, pp. 75–90.

Denford, S. 1978. Mites and their potential use in archaeology. In Brothwell, D. R., Thomas, K. D., and Clutton-Brock, J. (eds.), *Research Problems in Zooarchaeology. University of London, Institute of Archaeology Occasional Paper* 3:77–83.

Dévai, G., and Moldován, J. 1983. An attempt to trace eutrophication in a shallow lake (Balaton, Hungary) using chironomids. *Hydrobiologia* 103:169–175.

Dickson, J. H., Dickson, C. A., and Breeze, D. J. 1979. Flour or bread in a Roman military ditch at Bearsden, Scotland. *Antiquity* 53:47–51.

Dillehay, T. D. 1986. The cultural relationships of Monte Verde: A Late Pleistocene settlement site in the sub-antarctic forest of south-central Chile. In Bryan, A. L. (ed.), *New Evidence for the Pleistocene Peopling of the Americas.* Orono, Maine: Center for the Study of Early Man, pp. 319–337.

Dillehay, T. D. 1989. An evaluation of the paleoenvironmental reconstruction and the human presence at Monte Verde. In Dillehay, T. D. (ed.), *Monte Verde: A Late Pleistocene Settlement in Chile,* Volume 1: *Palaeoenvironment and Context.* Washington, D.C.: Smithsonian Institution Press, pp. 227–237.

Doyen, J. T., and Miller, S. E. 1980. Review of Pleistocene darkling ground beetles of the California asphalt deposits (Coleoptera: Tenebrionidae, Zopheridae). *Pan-Pacific Entomologist* 56:1–10.

Dredge, L. A., Morgan, A. V., and Nielsen, E. 1990. Sangamon and pre-Sangamon interglaciations in the Hudson Bay lowlands of Manitoba. *Géographie physique et Quaternaire* 44:319–336.

Dyke, A. S., and Matthews, J. V., Jr. 1987. Stratigraphy and paleoecology of Quaternary sediments along Pasley River, Boothia Peninsula, central Canadian arctic. *Géographie physique et Quaternaire* 41:323–344.

Dyke, A. S., and Prest, V. K. 1986. Late Wisconsin and Holocene retreat of the Laurentide ice sheet. Geological Survey of Canada Map, Scale 1:5,000,000, Map 1702A.

Egloff, M. 1987. 130 Years of archaeological research in Lake Neuchâtel, Switzerland. In Coles, J. M., and Lawson, A. J. (eds.), *European Wetlands in Prehistory.* Oxford: Clarendon Press, pp. 24–32.

Ehlers, J., Gibbard, P. L., and Rose, J. 1991. Glacial deposits of Britain and Europe: General overview. In Ehlers, J., Gibbard, P. L., and Rose, J. (eds.), *Glacial Deposits in Great Britain and Ireland.* Rotterdam, Netherlands: Balkema, pp. 493–501.

Elias, S. A. 1982a. Holocene insect fossils from two sites at Ennadai Lake, Keewatin, NWT. *Quaternary Research* 17:311–319.

Elias, S. A. 1982b. Bark beetle fossils from two early post-glacial age sites at high altitude in the Colorado Rocky Mountains. *Journal of Paleontology, Special Issue (Proceedings of the 3rd North American Paleontological Convention)* I:53–57.

Elias, S. A. 1982c. Holocene insect fossil assemblages from northeastern Labrador. *Arctic and Alpine Research* 14:311–319.

Elias, S. A. 1983. Paleoenvironmental interpretations of Holocene insect fossil assemblages from the La Poudre Pass site, northern Colorado Front Range. *Palaeogeography, Palaeoclimatology, Palaeoecology* 41:87–102.

Elias, S. A. 1985. Paleoenvironmental interpretations of Holocene insect fossil assemblages from four high-altitude sites in the Front Range, Colorado, U.S.A. *Arctic and
Alpine Research* 17:31–48.

Elias, S. A. 1986. Fossil insect evidence for Late Pleistocene paleoenvironments of the
Lamb Spring site, Colorado. *Geoarchaeology* 1:381–386.

Elias, S. A. 1987. Paleoenvironmental significance of Late Quaternary insect fossils
from packrat middens in south-central New Mexico. *Southwestern Naturalist*
32:383–390.

Elias, S. A. 1988a. Climatic significance of late Pleistocene insect fossils from Marias
Pass, Montana. *Canadian Journal of Earth Sciences* 25:922–926.

Elias, S. A. 1988b. Late Pleistocene paleoenvironmental studies from the Rocky
Mountain region: A comparison of pollen and insect fossil records. *Geoarchaeology*
3:147–153.

Elias, S. A. 1990a. Observations on the taphonomy of late Quaternary insect fossil
remains in packrat middens of the Chihuahuan Desert. *Palaios* 5:356–363.

Elias, S. A. 1990b. The timing and intensity of environmental changes during the
Paleoindian period in western North America: Evidence from the fossil insect
record. In Agenbroad, L. D., Mead, J. I., and Nelson, L. W. (eds.), *Megafauna and
Man*. Hot Springs, South Dakota, and Flagstaff, Arizona: Mammoth Site of Hot
Springs and Northern Arizona University, pp. 11–14.

Elias, S. A. 1991. Insects and climate change: Fossil evidence from the Rocky Mountains. *BioScience,* 41:552–559.

Elias, S. A. 1992a. Late Wisconsin insects and plant macrofossils associated with the
Colorado Creek mammoth, southwestern Alaska: Taphonomic and paleoenvironmental implications. In *22nd Arctic Workshop, Program and Abstracts,* pp. 45–47.

Elias, S. A. 1992b. Late Quaternary zoogeography of the Chihuahuan Desert insect
fauna, based on fossil records from packrat middens. *Journal of Biogeography*
19:285–298.

Elias, S. A. 1992c. Late Quaternary beetle fauna of southwestern Alaska: Evidence of a
refugium for mesic and hygrophilous species. *Arctic and Alpine Research* 24:133–144.

Elias, S. A. 1994. A paleoenvironmental setting for early Paleoindians in western
North America: Evidence from the insect fossil record. In Johnson, E. (ed.), *Ancient
Peoples and Landscapes*. Lubbock: Texas Tech University Press (in press).

Elias, S. A., and Halfpenny, J. C. 1991. Fox scat evidence of heavy predation on
beetles on the alpine tundra, Front Range, Colorado. *Coleopterists Bulletin*
45:189–190.

Elias, S. A., and Johnson, E. 1988. Pilot study of fossil beetles at the Lubbock Lake
Landmark. *Current Research in the Pleistocene* 5:57–59.

Elias, S. A., and Nelson, A. R. 1989. Fossil invertebrate evidence for Late Wisconsin environments at the Lamb Spring site, Colorado. *Plains Anthropologist*
34:309–326.

Elias, S. A., and Short, S. K. 1992. Paleoecology of an interglacial peat deposit,
Nuyakuk, southwestern Alaska. *Géographie physique et Quaternaire* 46:85–96.

Elias, S. A., and Toolin, L. J. 1989. Accelerator dating of a mixed assemblage of late Pleistocene insect fossils from the Lamb Spring site, Colorado. *Quaternary Research* 33:122–126.

Elias, S. A., and Van Devender, T. R. 1990. Fossil insect evidence for late Quaternary climatic change in the Big Bend region, Chihuahuan Desert, Texas. *Quaternary Research* 34:249–261.

Elias, S. A., and Van Devender, T. R. 1992. Insect fossil evidence of late Quaternary environments in the northern Chihuahuan Desert of Texas and New Mexico: Comparisons with the paleobotanical record. *Southwestern Naturalist* 37:101–116.

Elias, S. A., and Wilkinson, B. J. 1983. Lateglacial insect fossil assemblages from Lobsigensee (Swiss Plateau). Studies in the Late Quaternary of Lobsigensee 3. *Revue de Paléobiologie* 2:184–204.

Elias, S. A., Short, S. K., and Clark, P. U. 1986. Paleoenvironmental interpretations of the Late Holocene, Rocky Mountain National Park, Colorado, U.S.A. *Revue de Paléobiologie* 5:127–142.

Elias, S. A., Carrara, P. E., Toolin, L. J., and Jull, A. J. T. 1991. New radiocarbon and macrofossil analyses from Lake Emma, Colorado: Revision of the age of deglaciation. *Quaternary Research* 36:307–321.

Elias, S. A., Short, S. K., and Phillips, R. L. 1992a. Paleoecology of late glacial peats from the Bering Land Bridge, Chukchi Sea shelf region, northwestern Alaska. *Quaternary Research* 38:371–378.

Elias, S. A., Mead, J. I., and Agenbroad, L. D. 1992b. Late Quaternary arthropods from the Colorado Plateau, Arizona and Utah. *Great Basin Naturalist* 52:59–67.

Elias, S. A., Van Devender, T. R., and DeBaca, R. 1993. Insect fossil evidence of late glacial and Holocene environments in the Bolson de Mapimi, Chihuahuan Desert, Mexico: Comparisons with the paleobotanical record. *Palaios* (submitted).

Erickson, J. M. 1988. Fossil Oribatid mites as tools for Quaternary paleoecologists: preservation quality, quantities and taphonomy. In Laub, R. S., Miller, N. G., and Steadman, D. W. (eds.), *Late Pleistocene and Early Holocene Paleoecology and Archeology of the Eastern Great Lakes region. Bulletin of the Buffalo Society of Natural Sciences* 33:207–226.

Erochin, N. G., and Zinovjev, E. V. 1991. The Upper Pleistocene insect faunas from the localities of South and Middle Yamal. In *Ecological Groups of Ground Beetles (Coleoptera, Carabidae) in Natural and Anthropogenic Landscapes of the Ural.* Sverdlovsk: Ural Branch, USSR Academy of Sciences, pp. 18–22 [in Russian].

Erwin, T. L. 1979. Thoughts on the evolutionary history of ground beetles: Hypotheses generated from comparative faunal aanalyses of lowland forest sites in temperate and tropical regions. In Erwin, T. L., Ball, G. E., and Whitehead, D. R. (eds.), *Carabid Beetles: Their Evolution, Natural History, and Classification.* The Hague: Dr. W. Junk, pp. 539–592.

Erzinclioglu, Y. Z. 1983. The application of entomology to forensic medicine. *Medicine, Science, and the Law* 23:57–63.

Essig, E. O. 1927. Some insects from the adobe walls of the old missions of lower California. *Pan-Pacific Entomologist* 3:194–195.

Evans, W. G., and Baldwin, S. J. 1977. Larval exuviae of *Attagenus bicolor* Von Harold (Coleoptera: Dermestidae) from an archeological site at Mesa Verde, Colorado. *Quaestiones Entomologicae* 13:309–330.

Fairbanks, R. G. 1989. A 17,000-year glacio-eustatic sea level record: Influence of glacial melting rates on the Younger Dryas event and deep-ocean circulation. *Nature* 342:637–642.

Faulkner, C. T. 1991. Prehistoric diet and parasitic infection in Tennessee: Evidence from the analysis of desiccated human paleofeces. *American Antiquity* 56:687–700.

Fischer, R. C. 1988. An inordinate fondness for beetles. *Biological Journal of the Linnean Society* 35:313–319.

Fjellberg, A. 1978. Fragments of a Middle Weichselian fauna on Andøya, North Norway. *Boreas* 7:39.

Flach, K. 1884. Die Käfer der unterpleistocänen Ablagerungen bei Hösback, unweit Aschaffenburg. *Verhandlungen der Physikalisch-medizinischen Gesellschaft zu Würzburg (N.F.)* 18:285–297.

Fliche, P. 1875. Sur les lignites quaternaires de Jarville, près de Nancy. *Compte Rendu de l'Académie des Sciences, Paris* 80:1233–1236.

Fliche, P. 1876. Faune et flore des tourbières de la Champagne. *Compte Rendu de l'Académie des Sciences, Paris* 82:979–982.

Forman, S. L. 1989. Applications and limitations of thermoluminescence to date Quaternary sediments. *Quaternary International* 1:47–60.

Francoeur, A. 1983. The ant fauna near the tree-line in northern Quebec (Formicidae, Hymenoptera). *Nordicana* 47:177–180.

Francoeur, A., and Elias, S. A. 1985. *Dolichoderus taschenbergi* Mayr (Hymenoptera: Formicidae) from an early Holocene fossil insect assemblage in the Colorado Front Range. *Psyche* 92:303–308.

Franks, J. W., Sutcliffe, A., Kerney, M., and Coope, G. R. 1958. Haunt of elephant and rhinoceros: The Trafalgar Square of 100,000 years ago. *Illustrated London News,* June 14.

Fredskild, B. 1988. Agriculture in a marginal area—South Greenland from the Norse Landnam (985 A.D.) to the present (1985 A.D.). In Birks, H. B., Birks, H. J. B., Kaland, P. E., and Moe, D. (eds.), *The Cultural Landscape—Past, Present and Future.* Cambridge: Cambridge University Press, pp. 381–393.

Frey, D. G. 1986. Cladocera analysis. In Berglund, B. (ed.), *Handbook of Holocene Palaeoecology and Palaeohydrology.* New York: Wiley, pp. 667–692.

Fritz, P., Morgan, A. V., Eicher, U., and McAndrews, J. H. 1987. Stable isotopes, fossil Coleoptera and pollen stratigraphy in late Quaternary sediments from Ontario and New York State. *Palaeogeography, Palaeoclimatology, Palaeoecology* 58:183–202.

Froeschner, R. C. 1960. Cydnidae of the Western Hemisphere. *Proceedings of the United States National Museum* 111:337–680.

Fry, G. F. 1976. Analysis of prehistoric coprolites from Utah. *University of Utah Anthropological Papers* 97:1–45.

Futuyma, D. J. 1979. *Evolutionary Biology.* Sunderland, Massachusetts: Sinauer. 565 pp.

Gabus, J. H., Lemdahl, G., and Weidmann, M. 1987. Sur l'âge des terraces l'emanniques au SW de Lausanne. *Bulletin de Géologie Lausanne* 293:419–429.

Garry, C. E., Baker, R. G, Schwert, D. P., and Schneider, A. F. 1990a. Paleoenvironment of Twocreekan sediments at Kewaunee, Wisconsin as inferred from analyses of fossil beetle (Coleoptera) assemblages. In Schneider, A. F., and Fraser, G. S. (eds.), *Late Quaternary History of the Lake Michigan Basin. Geological Society of America Special Paper* 251:57–65.

Garry, C. E., Schwert, D. P., Baker, R. G., Kemmis, T. J., Horton, D. G., and Sullivan, A. E. 1990b. Plant and insect remains from the Wisconsinan interstadial/stadial transition at Wedron, north-central Illinois. *Quaternary Research* 33:387–399.

Gaunt, G. D., Coope, G. R., Osborne, P. J., and Franks, J. W. 1972. An interglacial deposit near Austerfield, southern Yorkshire. *Institute of Geological Sciences Report* 72/4.

Geyh, M. A., and Schleicher, H. 1990. Amino acid racemization method (AAR). In *Absolute Age Determination: Physical and Chemical Dating Methods and Their Application.* New York: Springer-Verlag, pp. 346–355.

Gibbard, P. L., Coope, G. R., Hall, A. R., Pierce, R. C., and Robinson, J. E. 1981. Middle Devensian deposits beneath the "Upper Floodplain" terrace of the River Thames at Kempton Park, Sunbury, England. *Proceedings of the Geological Association* 93:275–289.

Gilbert, M. B., and Bass, W. M. 1967. Seasonal dating of burials from the presence of fly pupae. *American Antiquity* 32:534–535.

Gillette, D. D., and Madsen, D. B. 1992. The short-faced bear *Arctodus simus* from the late Quaternary in the Wasatch Mountains of central Utah. *Journal of Vertebrate Paleontology* 12:107–112.

Giorgi, J., and Robinson, M. 1985. The environment. In Foreman, M., and Rahtz, S. (eds.), Excavations at Faccenda Chicken Farm, near Alchester, 1983. *Oxoniensia* 49:38–45.

Girling, M. A. 1973–1974. Fossil Coleoptera from Crossnacreevy. In Clayton, B., Palaeoecological investigations at Crossnacreevy, Co. Down. *Ulster Journal of Archaeology* 36–37:49–51.

Girling, M. A. 1974. Evidence from Lincolnshire of the age and intensity of the mid-Devensian temperate episode. *Nature* 250:270.

Girling, M. A. 1976a. Changes in the Meare Heath Coleoptera fauna in response to flooding. *Proceedings of the Prehistoric Society* 42:297–299.

Girling, M. A. 1976b. Fossil Coleoptera from the Somerset Levels: The Abbot's Way. *Somerset Levels Papers* 2:28–33.

Girling, M. A. 1977. Bird pellets from a Somerset Levels Neolithic trackway. *Naturalist (Hull)* 102:49–52.

Girling, M. A. 1978. The application of fossil insect studies to the Somerset Levels. In Brothwell, D. R., Thomas, K. D., and Clutton-Brock, J. (eds.), *Research Problems in Zooarchaeology. University of London, Institute of Archaeology Occasional Paper* 3:85–90.

Girling, M. A. 1979. The insects. In *Southwark Excavations 1972–1974. Joint Publication, London & Middlesex Archaeological Society and Surrey Archaeological Society, London* 1:170–172, 414, 466.

Girling, M. A. 1980. Two Late Pleistocene Insect Faunas from Lincolnshire. Ph.D. thesis, Department of Geology, University of Birmingham.

Girling, M. A. 1981. The environmental evidence. In Mellor, J. E., and Pearce, T., *The Austin Friars, Leicester. Council for British Archaeology, Research Report* 35:169–173.

Girling, M. A. 1982. The effect of the Meare Heath flooding episodes on the Coleopteran succession. *Somerset Levels Papers* 8:46–50.

Girling, M. A. 1983. The environmental implications of the excavations of 1974–76. In Brown, A. E., Woodfield, C., and Mynard, D. C., Excavations at Towcester, Northamptonshire: The Alchester Road suburb. *Northamptonshire Archaeology* 18:128–130 + fiche.

Girling, M. A. 1984. Aquatic Coleoptera in the fossil insect assemblages from archaeological sites in the Somerset Levels. *Balfour-Browne Club Newsletter* 30:1–11.

Girling, M. A. 1985a. The insect fauna—City Arms, trench 6, layer 3. In Shoesmith, R., Hereford City Excavations. 3. The Finds. *Council for British Archaeology Research Report* 56:96 + fiche M9.B12-14.

Girling, M. A. 1985b. Fly puparia. In Hirst, S. M., An Anglo-Saxon inhumation cemetery at Sewerby, East Yorkshire. *York University Archaeological Publications* 4:31 + fiche M1.A12.

Girling, M. A. 1986. The insects associated with Lindow Man. In Stead, I. M., Bourke, J. B., and Brothwell, D. (eds.), *Lindow Man: The Body in the Bog.* London: British Museum, pp. 90–91.

Girling, M. A. 1988. The bark beetle *Scolytus scolytus* (Fabricius) and the possible role of elm disease in the early Neolithic. In Jones, M. (ed.), Archaeology and the Flora of the British Isles. *Oxford University Committee for Archaeology Monograph* 14:34–38.

Girling, M. A. 1989. The insect fauna of the Roman Well at the Cattlemarket. *Chichester Excavations* 6:234–241.

Girling, M. A., and Greig, J. R. A. 1977. Palaeoecological investigations of a site at Hampstead Heath, London. *Nature* 268:45–47.

Girling, M. A., and Greig, J. R. A. 1985. A first fossil record for *Scolytus scolytus* (F.) (Elm Bark Beetle): Its occurrence in Elm Decline deposits from London and the implications for Neolithic Elm Disease. *Journal of Archaeological Science* 12:347–352.

Girling, M. A., and Robinson, M. 1987. The insect fauna. In Balaam, N. D., Bell, M. G., David, A. E. V., Levitan, B., McPhail, R. I., Robinson, M., and Scaife, R. G. (eds.), Prehistoric and Romano-British Sites at Westward Ho!, Devon. Archaeological and Palaeoenvironmental Surveys, 1983 and 1984. In Balaam, N. D., Levitan,

B., and Staker, V. (eds.), Studies in Palaeoeconomy and Environment in South West England. *British Archaeological Reports* 181:239–246.

Girling, M. A., and Robinson, M. 1989. Ecological interpretations from the beetle fauna. In Hayfield, C., and Greig, J. (eds.), Excavation and Salvage Work on a Moated Site at Cowick, South Humberside, 1976. *Yorkshire Archaeological Journal* 61:47–70.

Giterman, R. E., Sher, A. V., and Matthews, J. V., Jr. 1982. Comparison of the development of tundra-steppe environments in West and East Beringia: Pollen and macrofossil evidence from key sections. In Hopkins, D. M., Matthews, J. V., Jr., Schweger, C. E., and Young, S. B. (eds.), *Paleoecology of Beringia*. New York: Academic Press, pp. 43–73.

Golosova, L. D., Druk, A. Ya., Karppinen, E., and Kiselyov, S. V. 1985. Subfossil oribatid mites (Acarina, Oribatei) of northern Siberia. *Annales Entomologici Fennici* 51:3–18.

Gould, S. J., and Eldredge, N. 1977. Punctuated equilibria: The tempo and mode of evolution reconsidered. *Paleobiology* 3:115–151.

Goulet, H. 1983. The genera of Holarctic Elaphrini and species of *Elaphrus* Fabricius (Coleoptera: Carabidae): Classification, phylogeny and zoogeography. *Quaestiones Entomologicae* 19:219–482.

Graham, S. A. 1965. Entomology: An aid in archaeological studies. *American Antiquity Memoir* 19:167–174.

Green, C. P., Coope, G. R., Currant, A. P., Holyoak, D. T., Ivanovich, M., Jones, R. L., Keen, D. H., McGregor, D. F., and Robinson, J. E. 1984. Evidence of two temperate episodes in late Pleistocene deposits at Marsworth, UK. *Nature* 309:778–781.

Gregg, R. E. 1972. The northward distribution of ants in North America. *Canadian Entomologist* 104:1073–1091.

Greig, J. R. A. 1985. Agricultural diversity and sub-alpine colonisation: The story from pollen analysis at Fiave. In Malone, C., and Stoddart, S. (eds.), Papers in Italian Archaeology IV, Part 2: Prehistory. *British Archaeological Reports, International Series* 244:296–315.

Greig, J. R. A., Girling, M. A., and Skidmore, P. 1982. The plant and insect remains. In Barker, P., and Higham, R. (eds.), *Hen Domen, Montgomery: A timber castle on the English-Welsh Border.* London: Royal Archaeological Institute, pp. 60–71.

Grinnell, F. 1908. Quaternary myriapods and insects of California. *University of California Publications in Geology* 5:207–215.

Grunin, K. Ya. 1973. The first discovery of larvae of the mammoth bot-fly *Cobboldia* (*Mamontia,* subgen. N.) *russanovi* sp. n. (Diptera, Gasterophilidae). *Entomological Review* 52:228–230.

Günther, J. 1983. Development of Grossensee (Holstein, Germany): Variations in trophic status from the analysis of subfossil microfauna. *Hydrobiologia* 103:231–234.

Haarløv, N. 1967. Arthropoda (Acarina, Diptera) from subfossil layers in West Greenland. *Meddelelser om Grønland* 184:1–17.

Hadley, N. F. 1979. Wax secretion and color phases of the desert tenebrionid beetle *Cryptoglossa verrucosa* (LeConte). *Science* 203:367–369.

Haeck, J. 1971. The immigration and settlement of carabids in the new Ijselmeer-polders. *Miscellaneous Papers, Landbouwhogeschool, Wageningen* 8:33–53.

Hakbijl, T. 1987. Insect remains: Unadulterated Cantharidum and tobacco from the West Indies. In Gawronski, J. H. G. (ed.), *Amsterdam Project: Annual Report of the VOC Ship "Amsterdam" Foundation 1986.* Amsterdam, pp. 93–94.

Hakbijl, T. 1989. Insect remains from site Q, an Early Iron Age farmstead of the Assendelver Polders project. *Helinium* 29:77–102.

Hall, A. R., and Kenward, H. K. 1976. Biological evidence for the usage of Roman riverside warehouses at York. *Britannia* 7:274–276.

Hall, A. R., and Kenward, H. K. 1980. An interpretation of biological remains from Highgate, Beverley. *Journal of Archaeological Science* 7:33–51.

Hall, A. R., and Kenward, H. K. 1990. Environmental evidence from the Colonia. *Archaeology of York* 14/6.

Hall, A. R., Kenward, H. K., Williams, D., and Greig, J. R. A. 1983. Environment and living conditions at two Anglo-Scandinavian sites. *Archaeology of York,* 14/4.

Hall, H. J. 1977. A paleoscatalogical study of diet and disease at Dirty Shame Rock-shelter, southeast Oregon. *Tebiwa: Miscellaneous Papers of the Idaho State University Museum of Natural History* 8:1–14.

Hall, W. E., Van Devender, T. R., and Olson, C. A. 1988. Late Quaternary arthropod remains from Sonoran Desert packrat middens, southwestern Arizona and northwestern Sonora. *Quaternary Research* 29:277–293.

Hall, W. E., Olson, C. A., and Van Devender, T. R. 1989. Late Quaternary and modern arthropods from the Ajo Mountains of southwestern Arizona. *Pan-Pacific Entomologist* 65:322–347.

Hall, W. E., Van Devender, T. R., and Olson, C. A. 1990. Arthropod history of the Puerto Blanco Mountains, Organ Pipe National Monument, southwestern Arizona. In Betancourt, J. L., Van Devender, T. R., and Martin, P. S. (eds.), *Packrat Middens: The Last 40,000 Years of Biotic Change.* Tucson: University of Arizona Press, pp. 363–379.

Hamilton, T. D. 1991. The last interglaciation and the Old Crow tephra: Data from 10 Alaskan sites. In *21st Arctic Workshop, Program and Abstracts,* pp. 68–69.

Hammond, P., Morgan, A., and Morgan, A. V. 1979. On the *gibbulus* group of *Anotylus,* and fossil occurrence of *Anotylus gibbulus* (Staphylinidae). *Systematic Entomology* 4:215–221.

Hansen, J. P. H. 1989. The mummies from Qilakitsoq—paleopathological aspects. *Meddelelser om Grønland, Man and Society* 12:69–82.

Hebda, R. J., Warner, B. G., and Cannings, R. A. 1990. Pollen, plant macrofossils, and insects from fossil woodrat (*Neotoma cinerea*) middens in British Columbia. *Géographie physique et Quaternaire* 44:227–234.

Heer, O. 1865. *Die Urwelt der Schweiz.* 481 pp.

Heizer, R. F. 1970. The anthropology of Prehistoric Great Basin human coprolites. In Brothwell, D., and Higgs, E. (eds.), *Science in Archaeology*. New York: Praeger, pp. 244–250.

Helle, M., Sonstegaard, E., Coope, G. R., and Rye, N. 1981. Early Weichselian peat at Brumunddal, southeastern Norway. *Boreas* 10:369–379.

Hellqvist, M., and Lemdahl, G. 1990. Insektfynd från gamla Faulun. *Flora och Fauna* 85:234–239.

Henriksen, K. L. 1933. Undersøgelser over Danmark-Skånes kvartaere Insektfauna. *Videnskabelige Meddelelser fra Dansk Naturhistorisk Förening i København* 96: 77–355.

Heusser, C. 1989. Pollen analysis. In Dillehay, T. D. (ed.), *Monte Verde: A Late Pleistocene Settlement in Chile*, Volume 1: *Palaeoenvironment and Context*. Washington, D.C.: Smithsonian Institution Press, pp. 193–199.

Hevley, R. H., and Johnson, C. D. 1974. Insect remains from a Prehistoric Pueblo in Arizona. *Pan-Pacific Entomologist* 50:307–308.

Heydemann, B. 1967. Über die epigäische Aktivität terrestrischer Arthropoden der Küstenregionen im Tagerhythmus. In Graff, O., and Satchell, J. E. (eds.), *Progress in Soil Biology*. Braunschweig, Germany: Vieweg, pp. 249–263.

Hofmann, W. 1983. Stratigraphy of Cladocera and Chironomidae in a core from a shallow North German lake. *Hydrobiologia* 103:235–239.

Hofmann, W. 1986. Chironomid analysis. In Berglund, B. (ed.), *Handbook of Holocene Palaeoecology and Palaeohydrology*. New York: Wiley, pp. 715–727.

Hofmann, W. 1990. Weichselian chironomid and cladoceran assemblages from maar lakes. *Hydrobiologia* 214:207–212.

Hoganson, J. W., and Ashworth, A. C. 1992. Fossil beetle evidence for climatic change 18,000–10,000 years B.P. in south-central Chile. *Quaternary Research* 37:101–116.

Holdridge, L. 1987. The beetle remains. In Millett, M., and McGrail, S., The archaeology of the Hasholme Logboat. *Archaeological Journal* 144:88–89.

Holliday, V. T., Bozarth, S., Elias, S. A., Hall, S. A., Neck, R. W., and Winsborough, B. M. 1993. Stratigraphy and paleoenvironments of late Quaternary valley fills on the Southern High Plains. *Geological Society of America Memoir* (submitted).

Hopkins, D. M., Matthews, J. V., Jr., Wolfe, J. A., and Silberman, M. L. 1971. A Pliocene flora and insect fauna from the Bering Strait region. *Palaeogeography, Palaeoclimatology, Palaeoecology* 9:211–231.

Hopkins, D. M., Giterman, R. E., and Matthews, J. V., Jr. 1976. Interstadial mammoth remains and associated pollen and insect fossils, Kotzebue Sound area, northwestern Alaska. *Geology* 4:169–172.

Horn, G. H. 1876. Notes on some Coleopterous remains from the bone cave at Port Kennedy, Penna. *Transactions of the American Entomological Society* 5:241–245.

Horne, P. 1979. Head lice from an Aleutian mummy. *Paleopathology Newsletter* 25:7–8.

Howden, H. F. 1969. Effects of the Pleistocene on North American insects. *Annual Review of Entomology* 14:39–56.

Howden, H. F., and Cartwright, O. L. 1963. Scarab beetles of the genus *Onthophagus* Latreille north of Mexico (Coleoptera: Scarabaeidae). *Proceedings of the United States National Museum* 114(3467):1–135.

Howden, H. F., and Scholtz, C. H. 1986. Changes in a Texas dung beetle community between 1975 and 1985 (Coleoptera: Scarabaeidae, Scarabaeinae). *Coleopterists Bulletin* 40:313–316.

Hughes, O. L., Harrington, C. R., Janssens, J. A., Matthews, J. V., Jr., Morlan, R. E., Rutter, N. W., and Schweger, C. E. 1981. Upper Pleistocene stratigraphy, paleo-ecology, and archaeology of northern Yukon interior, eastern Beringia. 1. Bonnet Plume Basin. *Arctic* 34:329–365.

Hultén, E. 1937. *Outline of the History of Arctic and Boreal Biota during the Quaternary Period*. Lehre, Germany: J. Cramer. 168 pp.

Huntley, B., and Webb, T., III. 1989. Migration: Species' response to climatic variations caused by changes in the earth's orbit. *Journal of Biogeography* 16:5–19.

Iversen, J. 1954. The Late-Glacial flora of Denmark and its relation to climate and soil. *Danmarks Geologiska Undersökn* II(80).

Jones, A. K. G. 1976. The insect remains. In Rogerson, A., Excavations on Fuller's Hill, Great Yarmouth. *East Anglian Archaeology* 2:225.

Jónsson, J. 1892–1894. Faestebondens Kaar paa Island í det 18 Aarhundrede. *Dansk Historisk Tidsskrift* 6:563–604.

Jorgensen, G. 1986. Medieval plant remains from the settlements in Mollergade 6. In Jansen, H. M. (ed.), *The Archaeology of Svendborg, Denmark. 4. The Analysis of Medieval Plant Remains, Textiles and Wood from Svendborg*. Odense: Odense University Press, pp. 45–83.

Karppinen, E., and Koponen, M. 1974. Further observations on subfossil remains of oribatids (Acar., Oribatei) and insects in Piilonsuo, a bog in southern Finland. *Annales Entomologici Fennici* 40:172–175.

Kavanaugh, D. H. 1979. Rates of taxonomically significant differentiation in relation to geographical isolation and habitat: Examples from a study of the nearctic *Nebria* fauna. In Erwin, T. L., Ball, G. E., and Whitehead, D. R. (eds.), *Carabid Beetles: Their Evolution, Natural History, and Classification*. The Hague: Dr. W. Junk, pp. 37–57.

Keepax, C. A., Girling, M. A., Jones, R. T., Arthur, J. R. B., Paradine, P. J., and Keeley, H. 1979. The environmental analysis. In Williams, J. (ed.), *St. Peter's Square, Northampton, Excavations, 1973–1976*. Northampton: Northampton Development Corporation, p. 337.

Kelly, M., and Osborne, P. J. 1965. Two faunas and floras from the alluvium at Shustoke, Warwickshire. *Proceedings of the Linnean Society of London* 176:37–65.

Kenward, H. K. 1975. Pitfalls in the environmental interpretation of insect death assemblages. *Journal of Archaeological Science* 2:85–94.

Kenward, H. K. 1976. Reconstructing ancient ecological conditions from insects remains: Some problems and an experimental approach. *Ecological Entomology* 1:7–17.

Kenward, H. K. 1977. A note on the insect remains from Column Sample IV. In Armstrong, P., Excavations in Sewer Lane, Hull, 1974. *East Riding Archaeologist* 3:31–32.

Kenward, H. K. 1978. The value of insect remains as evidence of ecological conditions on archaeological sites. In Brothwell, D. R., Thomas, K. D., and Clutton-Brock, J. (eds.), *Research Problems in Zooarchaeology. Institute of Archaeology, University of London, Occasional Publication* 3:25–38.

Kenward, H. K. 1979a. Five insect assemblages. In Carver, M. O. H., Three Saxo-Norman tenements in Durham City. *Medieval Archaeology* 23:60–67.

Kenward, H. K. 1979b. The insect death assemblages. In Ayres, B. S., Excavations at Chapel Lane Staithe, 1978. Hull Old Town Reports Series 3. *East Riding Archaeologist* 5:65–72.

Kenward, H. K. 1980. Insect remains. In Schia, E. (ed.), Feltene "Oslogate 3 og 7." *De Arkeologiske Utgravninger i Gamlebyen, Oslo* 2:134–137.

Kenward, H. K. 1983. Beetles from Knap of Howe, Orkney. *Proceedings of the Society of Antiquaries of Scotland* 113:121.

Kenward, H. K. 1984. *Helophorus tuberculatus* Gyll. (Col., Hydrophilidae) from Roman Carlisle. *Entomologist's Monthly Magazine* 120:236.

Kenward, H. K. 1985a. The insect fauna. Berrington Street 4, period 6, pit 651. In Shoesmith, R., Hereford City Excavations. *Council for British Archaeology Research Report* 56:96 + fiche M9.B4-11.

Kenward, H. K. 1985b. Outdoors–indoors? The outdoor component of archaeological insect assemblages. In Fieller, N. R. J., Gilbertson, D. D., and Ralph, N. G. A. (eds.), Palaeobiological Investigations: Research Design, Methods and Data Analysis. *Association for Environmental Archaeology Symposium* 5B:97–104.

Kenward, H. K. 1988. *Helophorus tuberculatus* Gyllenhall (Col., Hydrophilidae) from Roman York. *Entomologist's Monthly Magazine* 124:90.

Kenward, H. K., and Allison, E. 1987. Waterlogged insect remains from Southampton. *Ancient Monuments Laboratory Report,* 124/87 (microfiche).

Kenward, H. K., and Girling, M. 1986. Arthropod remains from archaeological sites in Southampton. *Ancient Monuments Laboratory Report* 46/86 (microfiche).

Kenward, H. K., and Williams, D. 1979. Biological evidence from the Roman warehouses in Coney Street. *Archaeology of York* 14/2.

Kenward, H. K., Williams, D., Spencer, P. J., Greig, J. R. A., Rackham, D. J., and Brinklow, D. A. 1978. The environment of Anglo-Scandinavian York. In Hall, R. A. (ed.), Viking Age York and the North. *Council for British Archaeology Research Report* 27:58–70 .

Kenward, H. K., Hall, A. R., and Jones, A. K. G. 1986. Environmental evidence from a Roman well and Anglian pits in the Legionary Fortress. *Archaeology of York* 14/2.

Kenward, H. K., Engleman, C., Robertson, A., and Large, F. 1988. Insect remains from the Roman fort at Papcastle, Cumbria. *Ancient Monuments Laboratory Report* 145/88 (microfiche).

Kiselyov, S. V. 1973. Late Pleistocene Coleoptera of Transuralia. *Paleontological Journal* 4:507–510.

Kiselyov, S. V. 1988. Pleistocene and Holocene Coleoptera of western Siberia. In *Modern Condition and History of the Living World of the Western Siberian lowlands*. USSR Academy of Sciences, Ural Section, pp. 97–118 [in Russian].

Kiselyov, S. V., and Nazarov, V. I. 1984. Late Pleistocene insects. In Wright, H. E., Jr., and Barnowsky, C. W. (eds.), *Late Quaternary Environments of the Soviet Union*. Minneapolis: University of Minnesota Press, pp. 223–226.

Klinger, L. E., Elias, S. A., Behan-Pelletier, V. M., and Williams, N. E. 1990. The bog climax hypothesis: Fossil arthropod and stratigraphic evidence in peat sections from southeast Alaska, USA. *Holarctic Ecology* 13:72–80.

Klink, A. 1989. The Lower Rhine: Paleoecological analysis. In Petts, G. E. (ed.), *Historical Change of Large Alluvial Rivers: Western Europe*. London: Wiley, pp. 183–201.

Koch, K. 1970. Subfossile Käferreste aus römerzeitlichen und mittelalterlichen Ausgrabungen im Rheinland. *Entomologische Blätter für Biologie und Systematik der Käfer* 66:41–56.

Koch, K. 1971. Zur Untersuchung subfossiler Käferreste aus römerzeitlichen und mittelalterlichen Ausgrabungen im Rheinland. *Rheinische Ausgrabungen* 10:378–448.

Kolbe, H. 1894. Über fossile Reste von Coleopteren aus einem alten Torflager (Schmierkohle) bei Gr. Raschen in der nieder Lausitz. *Sitzberichte naturforschender Freunde zu Berlin*, 236–238.

Koponen, M., and Nuorteva, M. 1973. Über subfossile Waldinsekten aus dem Moor Piilonsuo in Südfinnland. *Acta Entomologica Fennica* 29:4–81.

Korotyayev, B. A. 1977. An ecological-faunistic survey of snout beetles (Coleoptera, Curculionidae) of the Northeast of the USSR. *Entomologicheskoe Obozrenie* 56:60–70 [in Russian].

Krivolutsky, D. A., and Druk, A. Ya. 1986. Fossil Oribatid mites. *Annual Review of Entomology* 31:533–545.

Krivolutsky, D. A., Druk, A. Ya., Eitminaviciute, I. S., Laskova, L. M., and Karppinen, E. 1990. *Fossil Oribatid Mites*. Vilnius, Lithuania: Mosklas. 109 pp. (in Russian).

Lea, P. D. 1989. Quaternary Environments and Depositional Systems of the Nushagak Lowlands, Southwestern Alaska. Ph.D. thesis, Department of Geological Sciences, University of Colorado, Boulder. 328 pp.

Lea, P. D., Elias, S. A., and Short, S. K. 1991. Stratigraphy and paleoenvironments of Pleistocene nonglacial units in the Nushagak Lowland, Southwestern Alaska. *Arctic and Alpine Research* 23:375–391.

LeConte, J. L. 1859. *The Coleoptera of Kansas and Eastern New Mexico*. Smithsonian Contributions to Knowledge, 1–6. 58 pp.

Leech, R., and Matthews, J. V., Jr. 1971. *Xysticus archaeopalpus* (Arachnidae: Thomisidae) A new species of crab spider from Pliocene sediments in western Alaska. *Canadian Entomologist* 103:1337–1340.

Lemdahl, G. 1982. Beetle remains from the refuse layer of the bogsite Ageröd V. In Larsson, L. (ed.), Ageröd V, an inland bogsite from the Atlantic period. *Acta Archaeologica Lundensia, Series* 12:169–172.

Lemdahl, G. 1985. Fossil insect faunas from Late-Glacial deposits in Scania (South-Sweden). *Ecologia Mediterranea* 11:185–191.

Lemdahl, G. 1988a. A postglacial insect fauna from Skateholm-Järavallen, southern Sweden. In L. Larsson (ed.), The Skateholm Project. I. Man and environment. *Acta Regiae Societatis Humanorum Litteratum Lundensis* 79:46–51.

Lemdahl, G. 1988b. Late Weichselian insect assemblages from the Kullen Peninsula, south Sweden—Palaeoenvironmental interpretations. *Lunqua Report* 30.

Lemdahl, G. 1988c. Insect remains from an Allerod peat at Lake Bysjön, south Sweden. *Boreas* 17:265–266.

Lemdahl, G. 1990. Insect assemblages from an Iron Age settlement in the clay district of Butjadingen, NW Germany. In *International Quaternary Union Subcommission for the Study of the Holocene, Cultural Landscapes Meetings Abstracts*, pp. 12–22.

Lemdahl, G. 1991a. A rapid climatic change at the end of the Younger Dryas in south Sweden—Paleoclimatic and paleoenvironmental reconstructions based on fossil insect assemblages. *Palaeogeography, Palaeoclimatology, Palaeoecology* 83:313–331.

Lemdahl, G. 1991b. Late Vistulian insect assemblages from Zabinko, western Poland. *Boreas* 20:71–77.

Lemdahl, G. 1991c. Tidigmedeltida insektfynd från Lund. *FaZett* 4:34–39.

Lemdahl, G. 1991d. Insetker. In Carlsson, R., Elfwendahl, M., and Perming, A. (eds.), Bryggaren ett kvarter i centrum—En medeltidsarkeologisk undersökning i Uppsala 1990. *Riksantikvarieämbetet och Statens Historiska Muséer Rapport UV (1991)* 1:222–223.

Lemdahl, G. 1991e. Insekter från Oskarshamnskoggen—Pilotundersökning av sedimentprover från vraket. *Uppsats i Påbyggnadskurs i arkeologi vid Stockholms Universitet* 1991:45–47.

Lemdahl, G., and Persson, T. 1989. Late Weichselian biostratigraphy at Mickelsmossen, Skåne, southern Sweden. *Geologioska Föreningens i Stockholm Förhandligar* 111:251–259.

Lemdahl, G., and Thelaus, M. 1989. Subfossila insektfynd från det medeltida Halmstad. *Entomologisk Tidskrift* 110:39–41.

Leonard, E. 1988. Estimates of Late-Pleistocene to Holocene climatic change in the Colorado Rocky Mountains. In *Tenth Biennial American Quaternary Association Meeting, Amherst, Massachusetts, Program & Abstracts*, p. 131.

Lindroth, C. H. 1948. Interglacial insect fossils from Sweden. *Arsbok Sveriges Geologiska Undersökning, Series C* 42:1–29.

Lindroth, C. H. 1949. Die Fennoskandischen Carabidae. *Kungliga Vetenskaps Vitterhets Samfundets Handlingar, Series B4* 1:1–709.

Lindroth, C. H. 1957. *The Faunal Connections Between North America and Europe.* New York: Wiley. 344 pp.

Lindroth, C. H. (ed.). 1960. *Catalogus Coleopterorum Fennoscandiae et Daniae.* Lund, Sweden: Entomologiska Sällskapet. 476 pp.

Lindroth, C. H. 1961. The ground beetles of Canada and Alaska, part 2. *Opuscula Entomologica, Supplement* 20:1–200.

Lindroth, C. H. 1963. The ground beetles of Canada and Alaska, part 3. *Opuscula Entomologica, Supplement* 24:201–408.

Lindroth, C. H. 1966. The ground beetles of Canada and Alaska, part 4. *Opuscula Entomologica, Supplement* 29:409–648.

Lindroth, C. H. 1968. The ground beetles of Canada and Alaska, part 5. *Opuscula Entomologica, Supplement* 33:649–944.

Lindroth, C. H. 1969. The ground beetles of Canada and Alaska, part 1. *Opuscula Entomologica, Supplement* 35:i–xlviii.

Lindroth, C. H. 1971. Biological investigations on the new volcanic island Surtsey, Iceland. In den Boer, P. J. (ed.), *Dispersal and Dispersal Power of Carabid Beetles. Miscellaneous Papers, Landbouwhogeschool, Wageningen* 8:65–69.

Lindroth, C. H. 1974. On elytral microsculpture of Carabid beetles. *Entomologica Scandinavica* 5:251–264.

Lindroth, C. H. 1985. The Carabidae (Coleoptera) of Fennoscandia and Denmark. *Fauna Entomologica Scandinavica* 15(1):1–227.

Lindroth, C. H. 1986. The Carabidae (Coleoptera) of Fennoscandia and Denmark. *Fauna Entomologica Scaninavica* 15(2):228–499.

Lindroth, C. H., and Coope, G. R. 1971. The insects from the interglacial deposits at Leveaniemi. *Sveriges Geologiska Undersökning, Series C* 658:44–55.

Lomnicki, A. M. 1894. Pleistocenskie owady z Borislawia. (Fauna Pleistocenica insectorum Boryslaviensium). *Wydawnictwa Muzeum imienia Dzieduszyckich* 4:1–116.

Lotter, A. 1991. Absolute dating of the Late-Glacial period in Switzerland using annually laminated sediments. *Quaternary Research* 35:321–330.

Lowell, T., Savage, K., Dell, A., Morgan, A. V., Pilny, J., Miller; N., Shane, L., and Stuckenrath, R. 1990. Late Wisconsin glacial maximum biota, Cincinnati, Ohio. In *Canadian Quaternary Association and American Quaternary Association, First Joint Meeting, Program and Abstracts*, p. 23.

Lundqvist, J. 1967. Submoräna sediment i Jämtlands län. *Sveriges Geologiska Undersökning* 618:1–267.

Lundqvist, J. 1986. Stratigraphy of the central area of the Scandinavian glaciation. *Quaternary Science Reviews* 5:251–268.

McCabe, A. M., Coope, G. R., Gennard, D. E., and Doughty, P. 1987. Freshwater organic deposits and stratified sediments between Early and Late Midlandian (Devensian) till sheets, at Aghnadarragh, County Antrim, Northern Ireland. *Journal of Quaternary Science* 2:11–33.

McGovern, T. H., Buckland, P. C., Savory, D., Sveinbjarnardóttir, G., Andreasen, C., and Skidmore, P. 1983. A study of the faunal and floral remains from two Norse farms in the western settlement, Greenland. *Arctic Anthropology* 20:93–120.

Mackay, W. P., and Elias, S. A. 1992. Late Quaternary ant fossils from Chihuahuan Desert packrat middens (Hymenoptera: Formicidae). *Psyche* 99:169–184.

McManus, D. A., and Creager, J. S. 1984. Sea-level data for parts of the Bering-Chukchi shelves of Beringia from 19,000 to 10,000 ^{14}C yr B.P. *Quaternary Research* 21:317–325.

McManus, D. A., Creager, J. S., Echols, R. J., and Holmes, M. L. 1983. The Holocene transgression of the arctic flank of Beringia: Chukchi Valley to Chukchi Estuary to Chukchi Sea. In Masters, P. M., and Flemming, M. C. (eds.), *Quaternary Coastlines and Marine Archeology*. New York: Academic Press, pp. 365–388.

Madsen, D. B., and Kirkman, J. E. 1988. Hunting hoppers. *American Antiquity* 53:593–604.

Mangerud, J. 1989. Correlation of the Eemian and Weichselian with deep sea oxygen isotope stratigraphy. *Quaternary International* 3/4:1–4.

Mani, M. S. 1968. *Ecology and Biogeography of High Altitude Insects*. The Hague: Dr. W. Junk. 527 pp.

Markgraf, V. 1986. Plant inertia reassessed. *American Naturalist* 127:725–726.

Markgraf, V., and Bradbury, J. P. 1982. Holocene climatic history of South America. *Striae* 16:40–45.

Matthews, J. V., Jr. 1968. A paleoenvironmental analysis of three Late Pleistocene coleopterous assemblages from Fairbanks, Alaska. *Quaestiones Entomologicae* 4:202–224.

Matthews, J. V., Jr. 1970. Two new species of *Micropeplus* from the Pliocene of western Alaska with remarks on the evolution of Micropeplinae (Coleoptera: Staphylinidae). *Canadian Journal of Zoology* 48:779–788.

Matthews, J. V., Jr. 1974a. Insect fossils from the early Pleistocene Olyor Suite (Chukochya River: Kolymian lowland, USSR). *Geological Survey of Canada Paper* 74-1A:207–211.

Matthews, J. V., Jr. 1974b. Quaternary environments at Cape Deceit (Seward Peninsula, Alaska): Evolution of a tundra ecosystem. *Geological Society of America Bulletin* 85:1353–1384.

Matthews, J. V., Jr. 1974c. Wisconsin environment of interior Alaska: Pollen and macrofossil analysis of a 26 meter core from the Isabella Basin (Fairbanks, Alaska). *Canadian Journal of Earth Sciences* 11:828–841.

Matthews, J. V., Jr. 1974d. A preliminary list of insect fossils from the Beaufort Formation, Meighen Island, District of Franklin. *Geological Survey of Canada, Papers* 74-1A:203–206.

Matthews, J. V., Jr. 1975a. Incongruence of macrofossil and pollen evidence: A case from the Late Pleistocene of the northern Yukon coast. *Geological Survey of Canada Paper* 75–1B:139–146.

Matthews, J. V., Jr. 1975b. Insects and plant macrofossils from two Quaternary exposures in the Old Crow–Porcupine region, Yukon Territory, Canada. *Arctic and Alpine Research* 7:249–259.

Matthews, J. V., Jr. 1976a. Evolution of the subgenus *Cyphelophorus* (Genus *Helophorus,* Hydrophilidae): Description of two new fossil species and discussion of *Helophorus tuberculatus* Gyll. *Canadian Journal of Zoology* 54:653–673.

Matthews, J. V., Jr. 1976b. Insect fossils from the Beaufort Formation: Geological and biological significance. *Geological Survey of Canada Papers* 76-1B:217–227.

Matthews, J. V., Jr. 1977a. Tertiary Coleoptera fossils from the North American arctic. *Coleopterists Bulletin* 31:297–308.

Matthews, J. V., Jr. 1977b. Coleoptera fossils: Their potential value for dating and correlation of late Cenozoic sediments. *Canadian Journal of Earth Sciences* 14:2339–2347.

Matthews, J. V., Jr. 1979a. Fossil beetles and the Late Cenozoic history of the tundra environment. In Gray, J., and Boucot, A. J. (eds.), *Historical Biogeography, Plate Tectonics, and the Changing Environment.* Corvallis: Oregon State University Press, pp. 371–378.

Matthews, J. V., Jr. 1979b. Tertiary and Quaternary environments: Historical background for an analysis of the Canadian insect fauna. In Danks, H. V. (ed.), Canada and its insect fauna. *Memoirs of the Entomological Society of Canada* 108:31–86.

Matthews, J. V., Jr. 1980a. Tertiary land bridges and their climate: Backdrop for development of the present Canadian insect fauna. *Canadian Entomologist* 112:1089–1103.

Matthews, J. V., Jr. 1980b. Paleoecology of John Klondike Bog, Fisherman Lake region, Southwest District of Mackenzie. *Geological Survey of Canada Paper* 80-22. 12 pp.

Matthews, J. V., Jr. 1981. Tertiary land bridges and their climate: Backdrop for the development of the present Canadian insect fauna. *Canadian Entomologist* 112:1089–1103.

Matthews, J. V., Jr. 1982. East Beringia during Late Wisconsin time: A review of the biotic evidence. In Hopkins, D. M., Matthews, J. V., Jr., Schweger, C. E., and Young, S. B. (eds.), *Paleoecology of Beringia.* New York: Academic Press, pp. 127–150.

Matthews, J. V., Jr. 1983. A method for comparison of northern fossil insect assemblages. *Géographie physique et Quaternaire* 37:297–306.

Matthews, J. V., Jr. 1989a. Late Tertiary arctic environments: A vision of the future? *Geos* 18:14–18.

Matthews, J. V., Jr. 1989b. Plants and other fossils from a late Tertiary beaver pond, Ellesmere Island, N.W.T. In *Paleobotany in Canada: Status and Prospects.* Ottawa: Canadian Museum of Nature. 2 pp.

Matthews, J. V., Jr., Mott, R. J., and Vincent, J.-S. 1986. Preglacial and interglacial environments of Banks Island: Pollen and macrofossils from Duck Hawk Bluffs and related sites. *Géographie physique et Quaternaire* 40:279–298.

Matthews, J. V., Jr., Smith, S. L., and Mott, R. J. 1987. Plant macrofossils, pollen, and insects of arctic affinity from Wisconsinan sediments in Chaudière Valley, southern Quebec. *Geological Survey of Canada Papers, Current Research, Series A* 87-1A:165–175.

Matthews, J. V., Jr., Schweger, C. E., and Janssens, J. A. 1990a. The Last (Koy-Yukon) interglaciation in the northern Yukon: Evidence from Unit 4 at Ch'ijee's Bluff, Bluefish Basin. *Géographie physique et Quaternaire* 44:341–362.

Matthews, J. V., Jr., Schweger, C. E., and Hughes, O. L. 1990b. Plant and insect fossils from the Mayo Indian Village section (Central Yukon): New data on middle Wisconsinan environments and glaciation. *Géographie physique et Quaternaire* 44:15–26.

Matthews, J. V., Jr., Ovenden, L. E., and Fyles, J. G. 1990c. Plant and insect fossils from the late Tertiary Beaufort Formation on Prince Patrick Island, N.W.T. In C. R. Harrington (ed.), *Canada's Missing Dimension: Science and History in the Canadian Arctic Islands*, Volume 1. Ottawa: National Museums of Canada, pp. 105–139

Mayr, E. 1970. *Populations, Species, and Evolution*. Cambridge, Massachusetts: Harvard University Press. 453 pp.

Milankovitch, M. 1941. *Kanon der Erdbestrahlung und seine Anwendung auf das Eiszeitproblem*. Belgrade, Yugoslavia: Académie Royale Serbe. 633 pp.

Miller, G. H., and Brigham-Grette, J. 1989. Amino acid geochronology: Resolution and precision in carbonate fossils. *Quaternary International* 1:111–128.

Miller, R. F. 1990. Coleoptera from the Sangamon Interglacial (?) in northern Ontario. In *Canadian Quaternary Association and American Quaternary Association, First Joint Meeting, Program and Abstracts*, p. 26.

Miller, R. F. 1991. Chitin paleoecology. *Biochemical Systematics and Ecology* 19:401–411.

Miller, R. F., and Morgan, A. V. 1982. A Postglacial coleopterous assemblage from Lockport Gulf, New York. *Quaternary Research* 17:258–274.

Miller, R. F., Morgan, A. V., and Hicock, S. R. 1985a. Pre-Vashon fossil Coleoptera of Fraser age from the Fraser Lowland, British Columbia. *Canadian Journal of Earth Sciences* 22:498–505.

Miller, R. F., Orr, G. L., Fritz, P., Downer, G., and Morgan, A. V. 1985b. Stable carbon isotope ratios in *Periplanata americana* L., the American cockroach. *Canadian Journal of Zoology* 63:584–589.

Miller, R. F., Fitzgerald, W. D., and Buhay, D. N. 1987. Fossil Coleoptera from the postglacial spruce-pine transition period near Minesing Swamp, Ontario. *Canadian Journal of Earth Sciences* 24:2099–2103.

Miller, R. F., Fritz, P., and Morgan, A. V. 1988. Climatic implications of D/H ratios in beetle chitin. *Palaeogeography, Palaeoclimatology, Palaeoecology* 66:277–288.

Miller, S. E. 1983. Late Quaternary insects of Rancho La Brea and McKittrick, California. *Quaternary Research* 20:90–104.

Miller, S. E., and Peck, S. B. 1979. Fossil carrion beetles of Pleistocene California asphalt deposits, with a synopsis of Holocene California Silphidae (Insecta: Coleoptera: Silphidae). *Transactions of the San Diego Society of Natural History* 19:85–106.

Miller, S. E., Gordon, R. D., and Howden, H. F. 1981. Reevaluation of Pleistocene scarab beetles from Rancho La Brea, California (Coleoptera: Scarabaeidae). *Proceedings of the Entomological Society of Washington* 83:625–630.

Mjöberg, E. 1904. Über eine schwedische interglaciale Coleopteren-species. *Geologiska Föreningens i Stockholm Förhandlingar* 26:493–497.

Mjöberg, E. 1905. Über eine schwedische interglaciale *Gyrinus*-Species. *Geologiska Föreningens i Stockholm Förhandlingar* 27:233–236.

Mjöberg, E. 1915. Über die Insektenreste der sogenannten "Härnögyttja" im nördlichen Schweden. *Sveriges Geologiska Undersokning* C268:1–14.

Mjöberg, E. 1916. Über die insektenreste der sog. "Härnögyttja" im nördlichen Schweden. *Geologiska föreningens i Stockholm Fördhandlingar* 38:222–223.

Moore, P. D. 1981. Life seen from a medieval latrine. *Nature* 294:614.

Moore, P. D. 1986. Bears versus beetles. *Nature* 320:385–386.

Morgan, A. 1969. A Pleistocene fauna and flora from Great Billing, Northampton-shire, England. *Opuscula Entomologica* 34:109–129.

Morgan, A. 1970. Weichselian Insect Faunas of the English Midlands. Ph.D. thesis, Department of Geology, University of Birmingham.

Morgan, A. 1972. The fossil occurrence of *Helophorus arcticus* Brown (Coleoptera Hydrophilidae) in Pleistocene deposits of the Scarborough Bluffs, Ontario. *Canadian Journal of Zoology* 50:555–558.

Morgan, A. 1973. Late Pleistocene environmental changes indicated by fossil insect faunas of the English Midlands. *Boreas* 2:173–212.

Morgan, A. 1975. Fossil beetle assemblages from the Early Wisconsinan Scarborough Formation. In *Quaternary Non-marine Paleoecology Conference, Abstracts,* p. 10.

Morgan, A., Morgan, A. V., and French, H. M. 1982. A Late Quaternary insect fauna from Belchatów, central Poland. *XI INQUA Congress, Moscow, Abstracts* 2:197.

Morgan, A., Morgan, A. V., and Elias, S. A. 1985. Holocene insects and palaeo-ecology of the Au Sable River, Michigan. *Ecology* 66:1817–1828.

Morgan, A. V. 1987. Late Wisconsin and early Holocene paleoenvironments of east-central North America based on assemblages of fossil Coleoptera. In Ruddiman, W. F., and Wright, H. E., Jr. (eds.), *The Geology of North America,* Volume K-3: *North America and Adjacent Oceans during the Last Deglaciation.* Boulder, Colorado: Geological Society of America, pp. 353–370.

Morgan, A. V. 1988. Late Pleistocene and early Holocene Coleoptera in the Lower Great Lakes region. In Laub, R. S., Miller, N. G., and Steadman, D. W. (eds.), *Late Pleistocene and Early Holocene Paleoecology and Archeology of the Eastern Great Lakes Region. Bulletin of the Buffalo Society of Natural Sciences* 33:195–206.

Morgan, A. V. 1989. Late Pleistocene zoogeographic shifts and new collecting records for *Helophorus arcticus* Brown (Coleoptera: Hydrophilidae) in North America. *Canadian Journal of Zoology* 67:1171–1179.

Morgan, A. V., and Freitag, R. 1982. The occurrence of *Cicindela limbalis* Klug (Coleoptera: Cicindelidae) in a late glacial site at Brampton, Ontario. *Coleopterists Bulletin* 36:105–108.

Morgan, A. V., and Morgan, A. 1976. Climatic interpretations from the fossil faunas of the Don and Scarborough Formations. *Geological Society of America, Annual Meeting, Abstracts* 8:1020.

Morgan, A. V., and Morgan, A. 1979. The fossil Coleoptera of the Two Creeks Forest Bed, Wisconsin. *Quaternary Research* 12:226–240.

Morgan, A. V., and Morgan, A. 1980. Faunal assemblages and distributional shifts of Coleoptera during the Late Pleistocene in Canada and the northern United States. *Canadian Entomologist* 112:1105–1128.

Morgan, A. V., and Morgan, A. 1986. A preliminary note on fossil insect faunas from central Illinois. *Quaternary Records of Central and North Illinois* 1:81–87.

Morgan, A. V., and Morgan, A. 1987. Paleoentomology—Towards the next decade. *Episodes* 10:38–40.

Morgan, A. V., Morgan, A., and Carter, L. D. 1979. Paleoenvironmental interpretation of a fossil insect fauna from bluffs along the lower Colville River, Alaska. In Johnson, K. M., and Williams, J. R. (eds.), The United States Geological Survey in Alaska: Accomplishments during 1978. *U.S. Geological Survey Circular* 804-B:41–44.

Morgan, A. V., Elias, S. A., and Morgan, A. 1981. Paleoenvironmental implications of a late glacial insect assemblage from southeast Michigan. In *Geological Society of Canada Annual Meeting, Calgary, Alberta, Abstracts*, p. A-41.

Morgan, A. V., Elias, S. A., and Morgan, A. 1982. Insect fossils from a late glacial site at Longswamp, Pennsylvania. In *Geological Association of Canada Annual Meeting, Calgary, Alberta, Abstracts*, p. A-41.

Morgan, A. V., McAndrews, J. M., Pilny, J. J., Goodwin, A. G., and Morgan, A. 1983a. Paleoecological reconstruction at the Rostock mammoth site, Ontario. *Geological Association of Canada, Program with Abstracts* 8A:48.

Morgan, A. V., Morgan, A., and Miller, R. F. 1983b. A preliminary report on two fossil insect assemblages from west-central Indiana. In Bleuer, N. K., Melhorn, W. N., and Pavey, R. R. (eds.), *Interlobate Stratigraphy of the Wabash Valley, Indiana.* Lafayette, Indiana: Purdue University Press, pp. 133–136.

Morgan, A. V., Morgan, A., Ashworth, A. C., and Matthews, J. V., Jr. 1984a. Late Wisconsin fossil beetles in North America. In Porter, S. C. (ed.), *Late Quaternary Environments of the United States,* Volume 1: *The Late Pleistocene.* Boulder, Colorado: Geological Society of America, pp. 354–363.

Morgan, A. V., Morgan, A., and Miller, R. F. 1984b. Range extension and fossil occurrences of *Holoboreaphilus nordenskioeldi* (Mäklin) (Coleoptera: Staphylinidae) in North America. *Canadian Journal of Zoology* 62:463–467.

Morlan, R. E., and Matthews, J. V., Jr. 1978. New dates for early man. *Geos 2–5.*

Morlan, R. E., and Matthews, J. V., Jr. 1983. Taphonomy and paleoecology of fossil insect assemblages from Old Crow River (CRH-15) northern Yukon Territory, Canada. *Géographie physique et Quaternaire* 37:147–157.

Morris, D. P. 1986. Human coprolites. In *Archaeological Investigations at Antelope House.* Washington, D.C.: National Park Service, pp. 165–188.

Mott, R. J. 1990. Sangamonian forest history and climate in Atlantic Canada. *Géographie physique et Quaternaire* 44:257–270.

Mott, R. J., and DiLabio, R. N. W. 1990. Paleoecology of organic deposits of probable last interglacial age in northern Ontario. *Géographie physique et Quaternaire* 44:309–318.

Mott, R. J., and Matthews, J. V., Jr. 1990. The Last (?) interglaciation in Canada. *Géographie physique et Quaternaire,* 44:245–248.

Mott, R. J., Anderson, T. W., and Matthews, J. V., Jr. 1981. Late-Glacial paleoenvironments of sites bordering the Champlain Sea based on pollen and macrofossil evidence. In Mahany, W. C. (ed.), *Quaternary Paleoclimate.* Norwich, England: Geoabstracts, pp. 129–172.

Mott, R. J., Anderson, T. W., and Matthews, J. V., Jr. 1982. Pollen and macrofossil study of an interglacial deposit in Nova Scotia. *Géographie physique et Quaternaire* 36:197–208.

Mott, R. J., Matthews, J. V., Jr., Grant, D. R., and Beke, G. J. 1986. A late glacial buried organic profile near Brookside, Nova Scotia. *Current Research, Part B, Geological Survey of Canada Paper* 86-1B:289–294.

Musson, C. R., Smith, A. G., and Girling, M. A. 1977. Environmental evidence from the Breiddin hillfort. *Antiquity* 51:147–151.

Nazarov, V. I. 1979. Coleoptera from the Rubezhnitsa locality and their environment. *Paleontological Journal* 13:462–470.

Nazarov, V. I. 1984. *A reconstruction of the Anthropogene landscapes in the northeastern part of Byelorussia according to palaeoentomological data.* Works of the Paleontological Institute. Moscow: Nauka Press. 95 pp [in Russian].

Nelson, R. E. 1982. Late Quaternary Environments of the Western Arctic Slope, Alaska. Ph. D. Thesis, University of Washington, Seattle.

Nelson, R. E. 1986. Mid-Wisconsin aridity in northern Alaska: Incongruity of a mesic pollen record and xeric insect indicators. *Current Research in the Pleistocene* 3:57–59.

Nelson, R. E. 1987. A postglacial flora and subfossil insect fauna from central Maine. In *Twelfth International Congress, International Union for Quaternary Research (INQUA), Ottawa, Ontario, Programme with Abstracts*, p. 231.

Nelson, R. E., and Carter, L. D. 1987. Paleoenvironmental analysis of insects and extralimital *Populus* from an early Holocene site on the Arctic Slope of Alaska. *Arctic and Alpine Research* 19:230–241.

Nelson, R. E., and Coope, G. R. 1982. A Late-Pleistocene insect fauna from Seattle, Washington. In *American Quaternary Association, Seventh Biennial Meeting, Seattle, Abstracts and Program*, p. 146.

Nichols, H. 1975. *Palynological and Paleoclimatic Study of the Late Quaternary Displacement of the Boreal Forest-Tundra Ecotone in Keewatin and Mackenzie, N.W.T., Canada.* University of Colorado, Institute of Arctic and Alpine Research, Occasional Paper 15. 87 pp.

Nielsen, E., Morgan, A. V., Morgan, A., Mott, R. J., Rutter, N. W., and Causse, C. 1986. Stratigraphy, paleoecology, and glacial history of the Gillam area, Manitoba. *Canadian Journal of Earth Sciences* 23:1641–1661.

Nissen, K. 1973. Analysis of human coprolites from Bamert Cave, Amador County, California. In Heizer, R. F., and Hester, T. R. (eds.), *The Archaeology of Bamert Cave, Amadore County, California.* Berkeley: University of California Archaeological Facility, pp. 65–71.

Noonan, G. R. 1985. The influences of dispersal, vicariance, and refugia on patterns of biogeographical distributions of the beetle family Carabidae. In Ball, G. E. (ed.), *Taxonomy, Phylogeny and Zoogeography of Beetles and Ants.* The Hague, Netherlands: Dr. W. Junk, pp. 322–350.

Noonan, G. R. 1988. Biogeography of North American and Mexican insects, and a critique of vicariance biogeography. *Systematic Zoology* 37:366–384.

Noonan, G. R. 1990. Biogeographical patterns of North American *Harpalus* Latreille (Insecta: Coleoptera: Carabidae). *Journal of Biogeography* 17:583–614.

O'Connor, J. P. 1979. *Blaps lethifera* Marsham (Coleoptera: Tenebrionidae), a beetle new to Ireland from Viking Dublin. *Entomologist's Gazette* 30:295–297.

O'Connor, J. P. 1987. Appendix V. Insecta. In Mitchell, G. F., Archaeology & Environment in Early Dublin. *Medieval Dublin Excavations* C1:39–40.

O'Connor, J. P., Hall, A. R., Jones, A. K. G., and Kenward, H. K. 1984. Ten years of environmental archaeology at York. In Addyman, P. V., and Black, V. E. (eds.), *Archaeological Papers from York Presented to M. W. Barley*. York, England: York Archaeological Trust, pp. 166–172.

Osborne, P. J. 1969. An insect fauna of Late Bronze Age date from Wilsford, Wiltshire. *Journal of Animal Ecology* 38:555–566.

Osborne, P. J. 1971a. An insect fauna from the Roman site at Alcester, Warwickshire. *Britannia* 2:156–165.

Osborne, P. J. 1971b. The insect fauna from the Roman harbour. In Cuncliffe, B. W., Excavations at Fishbourne, 1961–1969. *Research Reports of the Society of Antiquaries of London* 27:393–396.

Osborne, P. J. 1972. Insect faunas of Late Devensian and Flandrian age from Church Stretton, Shropshire. *Philosophical Transactions of the Royal Society of London, Series B* 263:327–367.

Osborne, P. J. 1973. Insects in archaeological deposits. *Science and Archaeology* 10:4–6.

Osborne, P. J. 1974. An insect assemblage of Early Flandrian age from Lea Marston, Warwickshire, and its bearing on contemporary climate and ecology. *Quaternary Research* 4:471–486.

Osborne, P. J. 1975. The Coleoptera from the Roman Well (on the intervallum road, east of the gyrus). In Hobley, B., The Lunt Roman Fort and Training School for Cavalry, Bagington, Warwickshire. *Transactions of the Birmingham and Warwickshire Archaeological Society* 87:44–46.

Osborne, P. J. 1976. The insect fauna of the peat within the lake deposits. In B. B. Clarke, Quaternary features of Porth Meare Cove, North Cornwall. *Transactions of the Royal Geological Society of Cornwall* 20:283–285.

Osborne, P. J. 1977. Stored product beetles from a Roman site at Droitwich, England. *Journal of Stored Products Research* 13:203–204.

Osborne, P. J. 1979. Insect remains. In Smith, C. (ed.), Fisherwick: The reconstruction of an Iron Age landscape. *British Archaeological Reports* 61:85–87, 189–193.

Osborne, P. J. 1980a. The insect fauna of the organic deposits at Sugworth and its environmental and stratigraphic implications. *Philosophical Transactions of the Royal Society of London, Series B* 289:119–133.

Osborne, P. J. 1980b. The Late Devensian–Flandrian transition depicted by serial insect faunas from West Bromwich, Staffordshire, England. *Boreas* 9:139–147.

Osborne, P. J. 1981a. Report of the insects from two organic deposits on the Smithfield Market site, Birmingham. In Watts, L., Birmingham Moat: Its history, topography, and destruction. *Transactions of the Birmingham and Warwickshire Archaeological Society* 89:63–66.

Osborne, P. J. 1981b. The insect fauna. In Stanford, S. C., *Midsummer Hill: An Iron Age Hillfort on the Malverns.* Hereford, England, pp. 156–157.

Osborne, P. J. 1981c. The insect fauna. In Jarrett, M. G., and Wrathmell, S., *Whitton: An Iron Age and Roman Farmstead in South Glamorgan.* Cardiff: University of Wales Press, pp. 245–248.

Osborne, P. J. 1981d. Coleopterous fauna from Layer 1. In Greig, J. R. A., The investigation of a Medieval barrel latrine from Worcester. *Journal of Archaeological Science* 8:268–271.

Osborne, P. J. 1982. Report on the insect remains from Cirencester. In Wacher, J., and McWhirr, A. (eds.), *Cirencester Excavations,* Volume 1: *Early Roman Occupation at Cirencester.* Cirencester, England: Cirencester Excavation Committee, p. 232.

Osborne, P. J. 1983. An insect fauna from a modern cesspit and its comparison with probable cesspit assemblages from archaeological Sites. *Journal of Archaeological Science* 10:453–463.

Osborne, P. J. 1984. Insect remains. In Leach, P. (ed.), The Archaeology of Taunton Excavations and Fieldwork to 1980. *Western Archaeological Trust, Excavation Monograph* 8:165–166.

Osborne, P. J. 1985. The insect remains from Fiave F4, 170cm. In Malone, C., and Stoddart, S. (eds.), Papers in Italian Archaeology IV, Part ii: Prehistory. *British Archaeological Reports* S244:308, 311–312.

Osborne, P. J. 1986. The Wilsford Shaft: Report on the insect fauna. *Ancient Monuments Laboratory Report* 8/86 (microfiche).

Osborne, P. J. 1989. Insects. In Ashbee, P., Bell, M., and Proudfoot, E. (eds.), *Wilsford Shaft: Excavations 1960–62.* London: English Heritage, pp. 96–99 + fiche C1–7.

Panagiotakopulu, E., and Buckland, P. C. 1991. Insect pests of stored products from Late Bronze Age Santorini, Greece. *Journal of Stored Products Research* 27:179–184.

Peake, D. S., and Osborne, P. J. 1971. The Wandle Gravels in the vicinity of Croydon. *Proceedings and Transactions of the Croydon Natural History and Scientific Society* 14:145–175.

Pearson, R. G. 1967. The significance of the Colney Heath beetle fauna. *Entomologist's Monthly Magazine* 103:169–170.

Pennington, W. 1986. Lags in adjustment of vegetation to climate caused by the pace of soil development: Evidence from Britain. *Vegetatio* 67:105–118.

Penny, L. F., Coope, G. R., and Catt, J. A. 1969. The age and insect fauna of the Dimlington silts, east Yorkshire. *Nature* 224:65–67.

Perry, D. W., Buckland, P. C., and Snaesdottir, M. 1985. The application of numerical techniques to insect assemblages from the site of Stóraborg, Iceland. *Journal of Archaeological Science* 12:335–345.

Phillips, R. L., and Brouwers, E. M. 1990. Vibracore stratigraphy of the northeastern Chukchi Sea, Alaska. In *19th Arctic Workshop, Program and Abstracts,* pp. 63–64.

Pierce, W. D. 1946. Fossil arthropods of California. 11. Description of the dung beetles (Scarabaeidae) of the tar pits. *Bulletin of the Southern California Academy of Sciences* 45:119–131.

Pierce, W. D. 1948. Fossil arthropods of California. 16. The carabid genus *Elaphrus* in the asphalt deposits. *Bulletin of the Southern California Academy of Sciences* 47:53–54.

Pierce, W. D. 1949. Fossil arthropods of California. 17. The silphid burying beetles in the asphalt deposits. *Bulletin of the Southern California Academy of Sciences* 48:55–70.

Pierce, W. D. 1954a. Fossil arthropods of California. 19. The Tenebrionidae-Scaurinae of the asphalt deposits. *Bulletin of the Southern California Academy of Sciences* 53:93–98.

Pierce, W. D. 1954b. Fossil arthropods of California. 20. The Tenebrionidae-Coniontinae of the asphalt deposits. *Bulletin of the Southern California Academy of Sciences* 53:142–156.

Pierce, W. D. 1957. Insects. *Geological Society of America Memoirs* 67:943–952.

Pilny, J. J., and Morgan, A. V. 1987. Paleoentomology and paleoecology of a possible Sangamonian site near Innerkip, Ontario. *Quaternary Research* 28:157–174.

Pilny, J. J., Morgan, A. V., and Morgan, A. 1987. Paleoclimatic implications of late Wisconsinan insect assemblages from Rostock, southwestern Ontario. *Canadian Journal of Earth Sciences* 24:617–630.

Ponel, P., and Coope, G. R. 1990. Lateglacial and early Flandrian Coleoptera from La Taphanel, Massif Central, France: Climatic and ecological implications. *Journal of Quaternary Science* 5:235–249.

Ponel, P., and Gadbin, C. 1988. Contribution of palaeoentomology to an episode in the vegetational history of the Lac d'Issarlés region. *Comptes Rendus Hebdomadaires des Séances de l'Académie des Sciences* 307:755–758 (in French).

Ponel, P., deBeaulieu, J. L., and Tobolski, K. 1992. Holocene paleoenvironment at the timberline in the Taillefer Massif: Pollen analysis, study of plant and insect macrofossils. *The Holocene* 2:117–130.

Prentice, I. C. 1983. Postglacial climatic change: Vegetation dynamics and the pollen record. *Progress in Physical Geography* 7:273–286.

Prentice, I. C. 1986. Vegetation responses to past climatic variation. *Vegetatio* 67:131–141.

Prentice, I. C., Bartlein, P. J., and Webb, T., III. 1992. Vegetation and climate change in eastern North America since the last glacial maximum. *Ecology* 72:2038–2056.

Prest, V. K., Terasme, J., and Matthews, J. V., Jr. 1976. Late-Quaternary history of Magdalen Islands, Quebec. *Maritime Sediments* 12:39–59.

Price, P. W. 1984. *Insect Ecology.* New York: Wiley. 607 pp.

Ratcliffe, B. C., and Fagerstrom, J. A. 1980. Invertebrate lebensspuren of Holocene floodplains: Their morphology, origin and paleoecological significance. *Journal of Paleontology* 54:614–630.

Ritchie, J. C. 1986. Climate change and vegetation response. *Vegetatio* 67:65–74.

Roback, S. S. 1970. The Chironomidae. In Hutchinson, G. E. (ed.), An account of the history and development of the Lago di Monterosi, Latium, Italy. *Transactions of the American Philosophical Society* 60:150–162.

Robertsson, A.-M. 1988. Biostratigraphical studies of interglacial and interstadial deposits in Sweden. *University of Stockholm, Department of Quaternary Research Report* 10:1–19.

Robinson, M. A. 1975. The environment of the Roman defences at Alchester and its implications. *Oxoniensia* 40:161–170.

Robinson, M. A. 1978. Insect and seed remains. In Brodribb, A. C. C., Hands, A. R., and Walker, D. R., Excavations at Shakenoak Farm, near Wilcote, Oxfordshire, Part V: Sites K and E. Oxford: British Archaeological Reports, pp. 161–164.

Robinson, M. A. 1979a. The biological evidence. In Lambrick, G., and Robinson, M. A., Iron Age and Roman riverside settlements at Farmoor, Oxfordshire. *Council for British Archaeology Research Report* 32:77–133.

Robinson, M. A. 1979b. Calcified seeds and arthropods. In Parrington, M. (ed.), Excavations at Stert Street, Abingdon, Oxon. *Oxoniensia* 44:23–24.

Robinson, M. A. 1980. Appendix 10: Insect remains. In Christie, P. M., and Coad, J. G., Excavations at Denny Abbey. *Archaeological Journal* 137:267.

Robinson, M. A. 1981a. Roman waterlogged plant and invertebrate evidence. In Hinchliffe, J., and Thomas, R., Archaeological investigations at Appleford. *Oxoniensia* 45:90–106.

Robinson, M. A. 1981b. Waterlogged plant and invertebrate evidence. In Palmer, N., A beaker burial and medieval tenements in the Hamel, Oxford. *Oxoniensia* 45:199–206 + fiche.

Robinson, M. A. 1983. The insect remains. In Williams, J. H., and Farwell, D., Excavations of a Saxon Site in St. James' Square, Northampton, 1981. *Northamptonshire Archaeology* 18:Fiche M33 and M38.

Robinson, M. A. 1984. Plant and invertebrate remains. In Halpin, C., Late Saxon evidence and excavation of Hinxey Hall, Queen Street, Oxford. *Oxoniensia* 48:69 + fiche.

Robinson, M. 1986. Plant and invertebrate remains from the priory drains. In Lambrick, G., Further excavations on the second site of the Dominican Priory, Oxford. *Oxoniensia* 50:196–201 + fiche E10-F1.

Ruddiman, W. F., and McIntyre, A. 1981. The North Atlantic Ocean during the last deglaciation. *Palaeogeography, Palaeoclimatology, Palaeoecology* 35:145–214.

Sadler, J. P. 1990. Records of ectoparasites on humans and sheep from Viking Age deposits in the former Western Settlement on Greenland. *Journal of Medical Entomology* 27:628–631.

Salonen, V. P., Ikaheimo, M., and Luoto, J. 1981. The prehistoric settlement in the Kuppajarvi area of Pukko, S.W. Finland (Varsinais Suomi) in the light of palaeontology and archaeology. *Publications of the Department of Quaternary Geology, University of Turku* 44 (in Finnish).

Samuels, J., and Buckland, P. C. 1978. A Romano-British settlement at Sandtoft, South Humberside. *Yorkshire Archaeological Journal* 50:65–75.

Schaff, E. 1892. Über Insektenreste aus dem Torflager von Klinge. *Sitzungsberichte der Gesellschaft naturforschender Freunde zu Berlin.*

Schelvis, J. 1987. Some aspects of research on mites (Acari) in archaeological samples. *Palaeohistoria* 29:211–218.

Schelvis, J. 1990a. Mites from Medieval Oldeboorn. *Experimental and Applied Entomology, Proceedings of the Netherlands Entomological Society* 1:96–97.

Schelvis J. 1990b. The reconstruction of local environments on the basis of remains of oribatid mites (Acari; Oribatida). *Journal of Archaeological Science* 17:559–571.

Schweger, C. E., Matthews, J. V., Jr., Hopkins, D. M., and Young, S. B. 1982. Paleoecology of Beringia: A synthesis. In Hopkins, D. M., Matthews, J. V., Jr., Schweger, C. E., and Young, S. B. (eds.), *Paleoecology of Beringia*. New York: Academic Press, pp. 425–444.

Schweiger, H. 1967. Gebirgssysteme als Zentren der Artbildung. *Deutsche Entomologische Zeitschrift* 16:159–174.

Schwert, D. P. 1992. Faunal transitions in response to an ice age: The late Wisconsinan record of Coleoptera in the north-central United States. *Coleopterists Bulletin* 46:68–94.

Schwert, D. P., and Ashworth, A. C. 1985. Fossil evidence for late Holocene faunal stability in southern Minnesota (Coleoptera). *Coleopterists Bulletin* 39:67–79.

Schwert, D. P., and Ashworth, A. C. 1988. Late Quaternary history of the northern beetle fauna of North America: A synthesis of fossil distributional evidence. *Memoirs of the Entomological Society of Canada* 144:93–107.

Schwert, D. P., and Bajc, A. F. 1989. Fossil Coleoptera associated with the Moorhead Low-Water Phase of Lake Agassiz in northwestern Ontario. In *Geological Society of America Annual Meeting, Program and Abstracts,* p. 345.

Schwert, D. P., and Morgan, A. V. 1980. Paleoenvironmental implications of a late glacial insect assemblage from northwestern New York. *Quaternary Research* 13:93–110.

Schwert, D. P., Anderson, T. W., Morgan, A., Morgan, A. V., and Karrow, P. 1985. Changes in Late Quaternary vegetation and insect communities in southern Ontario. *Quaternary Research* 23:205–226.

Scudder, S. H. 1877. Description of two new species of Carabidae found in the Interglacial deposits at Scarborough Hts., in Toronto, Canada. *United States Geological Survey of the Territories (Hayden)* B3:763–764.

Scudder, S. H. 1890. The Tertiary insects of North America. *Report of the United States Geological Survey of the Territories* 13:13–90.

Scudder, S. H. 1898. The Pleistocene beetles of Fort River, Massachusetts. *Monographs of the United States Geological Survey* 24:740–746.

Scudder, S. H. 1900. Canadian fossil insects. *Contributions to Canadian Palaeontology* 2 (2):67–92.

Shackleton, N. J., and Opdyke, N. D. 1973. Oxygen isotope and palaeomagnetic stratigraphy of equatorial Pacific core V28-238: Oxygen isotope temperatures and ice volumes on a 10–5 year and 10–6 year scale. *Quaternary Research* 3:39–55.

Sher, A. V., Giterman, R. Ye., Zazhigan, V. S., and Kiselyov, S. V. 1977. New data on the Late Cenozoic sediments of the Kolyma lowland. *Proceedings of the USSR Academy of Sciences, Geology Series* 5:69–83 (in Russian).

Short, S. K. 1985. Palynology of Holocene sediments, Colorado Front Range: Vegetation and treeline changes in the subalpine forest. *American Association of Stratigraphic Palynologists Contribution Series* 16:7–30.

Short, S. K., and Elias, S. A. 1987. New pollen and beetle analysis at the Mary Jane site, Colorado: Evidence for Late-Glacial tundra conditions. *Geological Society of America Bulletin* 98:540–548.

Short, S. K., Elias, S. A., Waythomas, C. F., and Williams, N. E. 1992. Holocene environments, Nushagak and Holitna lowlands, southwestern Alaska. *Arctic* 45:381–392.

Shotton, F. W., and Coope, G. R. 1983. Exposures in the Power House Terrace of the River Stour at Wilden, Worcestershire, England. *Proceedings of the Geological Association* 94:33–44.

Shotton, F. W., and Osborne, P. J. 1965. The fauna of the Hoxnian Interglacial deposits of Nechells, Birmingham. *Philosophical Transactions of the Royal Society of London, Series B* 248:353–378.

Shotton, F. W., Sutcliffe, A. J., and West, R. G. 1962. The fauna and flora from the brick pit at Lexden, Essex. *Essex Naturalist* 31:15–22.

Shotton, F. W., Williams, R. E. G., and Johnson, A. S. 1974. Birmingham University radiocarbon dates VIII. *Radiocarbon* 16:285–303.

Shotton, F. W., Osborne, P. J., and Greig, J. R. A. 1977. The fossil content of a Flandrian deposit at Alcester. *Proceedings of the Coventry and District Natural History and Scientific Society* 5:19–32.

Skidmore P. 1986. The dipterous remains. In Stead, I. M., Bourke, J. B., and Brothwell, D. (eds.), *Lindow Man: The Body in the Bog*. London: British Museum, p. 92.

Smetana, A. 1985. Revision of the subfamily Helophorinae of the Nearctic region (Coleoptera: Hydrophilidae). *Memoirs of the Entomological Society of Canada* 131. 153 pp.

Smith, B. 1981. Insect remains. In Neil, N. R. J., A Bronze Age burial at Holland, St. Ola, Orkney. *Glasgow Archaeological Journal* 8:45.

Smith, C. 1978. The landscape and natural history of Iron Age settlement on the Trent Gravels. In Cunliffe, B. W., and Rowley, T. (eds.), Lowland Iron Age Communities in Europe. *British Archaeological Reports* S48:91–101.

Smith, D. 1989. Insect remains from site DSR. In Buckland, P. C., Magilton, J. R., and Hayfield, C., The Archaeology of Doncaster 2: The Medieval and Later Town. *British Archaeological Reports* 202:447–450.

Smith, K. G. V. 1986. *A Manual of Forensic Entomology*. London: British Museum (Natural History). 205 pp.

Solomon, M. E. 1979. Archaeological records of storage pests: *Sitophilus granarius* (L.) (Coleoptera, Curculionidae) from an Egyptian pyramid tomb. *Journal of Stored Product Research* 1:105–107.

Stanford, D., Wedel, W. R., and Scott, G. R. 1982. Archaeological investigations of the Lamb Spring site. *Southwestern Lore* 47:14–27.

Stanley, S. M. 1985. Rates of evolution. *Paleobiology* 11:13–26.

Stewart, J. H. 1938. Basin-plateau aboriginal sociopolitical groups. *Bureau of American Ethnology Bulletin* 33:3.

Stiger, M. A. 1977. Anasazi Diet: The Coprolite Evidence. Unpublished masters thesis, Department of Anthropology, University of Colorado, Boulder.

Strobel, P., and Pigorni, L. 1864. Le terremare e le palafitte del Parmense, seconda relazione. *Atti della Societa italiana di Scienze Naturali, Milan* 7:36–37.

Sveinbjarnardóttir, G. 1983. Palaeoekologiske undersogelser pa Holt i Eyjafjallasveit, Sydisland. In Olafsson, G. (ed.), *Hus, Gard och Bebyggelse. Föredrag fran det XVI nordiska arkeologmötet, Island, 1982* 4:241–250.

Sveinbjarnardóttir, G., and Buckland, P. C. 1983. An uninvited guest. *Antiquity* 57:127–130.

Taylor, B. J., and Coope, G. R. 1985. Arthropods in the Quaternary of East Anglia— Their role as indices of local paleoenvironments and regional paleoclimates. *Modern Geology* 9:159–185.

Ters, M., Azema, C., Brebion, P., Churcher, C. S., Delibrias, G., Denefle, M., Guyader, J., Lauriet, A., Mathieu, R., Michel, J. P., Osborne, P. J., Rouvillois, A., and Shotton, F. W. 1971. Sur le remblaiement holocène dans l'estuaire de la Seine, au Havre (Seine Maritime), France. *Quaternaria* 14:151–174.

Thiele, H. U. 1977. *Carabid Beetles in Their Environments.* New York: Springer-Verlag. 369 pp.

Tobolski, K. 1988. Palaeobotanical study of Bølling sediments of Zabinko in the vicinity of Poznan, Poland. *Quaestiones Geographicae* 10:119–124.

Totten, S. M. 1971. The occurrence of beetle remains in Pleistocene deposits, east-central United States. *Geological Society of America, Annual Meeting (Washington, D.C.), Abstracts with Program* 3:733–734.

Uutala, A. J. 1990. *Chaoborus* (Diptera: Chaoboridae) mandibles—Paleolimnological indicators of the historical status of fish populations in acid-sensitive lakes. *Journal of Paleolimnology* 4:139–151.

Van Dyke, E. C. 1939. The origin and distribution of the coleopterous insect fauna of North America. *Proceedings of the Sixth Pacific Science Congress* 4:255–268.

Van Geel, B., Coope, G. R., and Van der Hammen, T. 1989. Palaeoecology and stratigraphy of the lateglacial type section at Usselo (The Netherlands). *Review of Palaeobotany and Palynology* 60:25–129.

Velichko, A. A., and Faustova, M. A. 1986. Glaciations in the east European region of the USSR. *Quaternary Science Reviews* 5:447–461.

Vuilleumier, B. S. 1971. Pleistocene changes in the fauna and flora of South America. *Science* 173:771–780.

Waage, J. K. 1976. Insect remains from ground sloth dung. *Journal of Paleontology* 50:991–994.

Walker, I. R. 1987. Chironomidae (Diptera) in paleoecology. *Quaternary Science Reviews* 6:29–40.

Walker, I. R., and Mathewes, R. W. 1987. Chironomidae (Diptera) and postglacial climate at Marion Lake, British Columbia, Canada. *Quaternary Research* 27:89–102.

Walker, I. R., and Mathewes, R. W. 1988. Late-Quaternary fossil chironomidae (Diptera) from Hippa Lake, Queen Charlotte Islands, British Columbia, with special reference to *Corynocera* Zett. *Canadian Entomologist* 120:739–751.

Walker, I. R., and Mathewes, R. W. 1989. Early postglacial chironomid succession in southwestern British Columbia, Canada, and its paleoenvironmental significance. *Journal of Paleolimnology* 2:1–14.

Walker, I. R., and Paterson, C. G. 1983. Post-glacial chironomid succession in two small, humic lakes in the New Brunswick–Nova Scotia (Canada) border area. *Freshwater Invertebrate Biology* 2:61–73.

Walker, I., Mott, R. J., and Smol, J. P. 1991a. Alleröd–Younger Dryas lake temperatures from midge fossils in Atlantic Canada. *Science* 253:1010–1012.

Walker, I. R., Smol, J. P., Engstrom, D. R., and Birks, H. J. B. 1991b. An assessment of Chironomidae as quantitative indicators of past climatic change. *Canadian Journal of Fisheries and Aquatic Science* 48:975.

Walter, R. C. 1989. Application and limitation of fission-track geochronology to Quaternary tephras. *Quaternary International* 1:35–46.

Warner, B. G., Karrow, P. F., Morgan, A. V., and Morgan, A. 1987. Plant and insect fossils from Nipissing sediments along the Goulais River, southeastern Lake Superior. *Canadian Journal of Earth Sciences* 24:1526–1536.

Warner, B. G., Morgan, A. V., and Karrow, P. F. 1988. A Wisconsinan interstadial arctic flora and insect fauna from Clarksburg, southwestern Ontario, Canada. *Palaeoegeography, Palaeoclimatology, Palaeoecology* 68:27–47.

Watts, W. A. 1979. Late Quaternary vegetation of central Appalachia and the New Jersey Coastal Plain. *Ecological Monographs* 49:427–469.

Waythomas, C. F. 1990. Quaternary Geology and Late-Quaternary Environments of the Holitna Lowland, and Chuilnuk-Kioluk Mountains Region, Interior Southwestern Alaska. Ph.D. thesis, Department of Geological Sciences, University of Colorado, Boulder.

Waythomas, C. F., Short, S. K., and Elias, S. A. 1989. Deposits of interglacial character, Nenana Valley, Alaska. *Current Research in the Pleistocene* 6:95–98.

Waythomas, C. F., Short, S. K., and Elias, S. A. 1993. Paleoenvironmental study of late Quaternary organic sediments at the Foraker River, northern Alaska Range, Alaska. *Boreas* (submitted).

Wesenberg-Lund, C. 1896. Om Ferskvandsfaunaens Kitin-og Kisellemninger i Torvelagene. *Meddelelser fra Dansk geologisk Forening* 3:51–84.

Westgate, J. A. 1988. Isothermal plateau fission-track age of the Late Pleistocene Old Crow tephra, Alaska. *Geophysical Research Letters* 15:376–379.

White, R. E. 1983. *A Field Guide to the Beetles of North America*. Boston: Houghton Mifflin. 368 pp.

Wickham, H. F. 1917. Some fossil beetles from the Sangamon peat, Champaign Co., Ill. *American Journal of Science, 4th Series* 44:137–145.

Wickham, H. F. 1919. Fossil beetles from Vero, Florida. *American Journal of Science, 4th Series* 47:355–357.

Wiggins, G. B. 1977. *Larvae of the North American Caddisfly Genera (Trichoptera)*. Toronto: University of Toronto Press. 401 pp.

Williams, N. E. 1988. The use of caddisflies (Trichoptera) in paleoecology. *Palaeogeography, Palaeoclimatology, Palaeoecology* 62:493–500.

Williams, N. E., and Morgan, A. V. 1977. Fossil caddisflies (Insecta: Trichoptera) from the Don Formation, Toronto, Ontario, and their use in paleoecology. *Canadian Journal of Zoology* 55:519–527.

Williams, N. E., Westgate, J. A., Williams, D. D., and Morgan, A. V. 1981. Invertebrate fossils (Insecta: Trichoptera, Diptera, Coleoptera) from the Pleistocene Scarborough Formation at Toronto, Ontario and their paleoenvironmental significance. *Quaternary Research* 16:146–166.

Williams, N. E., Morgan, A. V., and Elias, S. A. 1993. Quaternary insect responses to environmental change in the North American Great Lakes region. *Quaternary Research* (submitted).

Wilson, E. C. 1986. Type specimens of fossil Invertebrata in the Natural History Museum of Los Angeles County: Supplement II. *Natural History Museum of Los Angeles County, Technical Report* 1:1–150.

Wilson, M. J., and Elias, S. A. 1986. Paleoecological significance of Holocene insect fossil assemblages from the North Coast of Alaska. *Arctic* 39:150–157.

Wintle, A. G., and Westgate, J. A. 1986. Thermoluminescence age of Old Crow tephra in Alaska. *Geology* 14:594–597.

Wollaston, T. V. 1863. Note on the remains of Coleoptera from the peat of Lexden Brick-pit. *Quarterly Journal of the Geological Society* 19:400–401.

Young, F. N. 1959. Fossil beetles from the Vero Pleistocene. *Coleopterists Bulletin* 13:103–106.

GLOSSARY

aedeagus   the male reproductive organ of insects.

Alleröd pollen zone   an interval of climatic amelioration during the late glacial, from 12,000 to 11,000 yr B.P., which favored the growth of birch, pine, and willow in Europe.

Altithermal   an interval of maximum warming in the Holocene, which is thought to have occurred from 9000 to 5000 yr B.P.

amino acid racemization   a method of dating fossils by measuring the extent to which the left-hand configuration of certain amino acids has racemized to the right-hand configuration. This process is used as a measure of time since the organism died.

anoxic sediment   sediment depleted of oxygen, in which decomposition is slowed or halted.

anthropogenic   of or pertaining to changes caused by humans.

archaeoentomology   the study of insect fossils from archaeological sites.

bedding plane   a planar bedding surface that visibly separates each successive layer of strata in a section.

Beringia   Eastern Siberia, Alaska, the Yukon Territory west of the Mackenzie River, and the continental shelf region between Siberia and Alaska. This region was largely unglaciated during the late Pleistocene and formed a high-latitude refugium for arctic biota.

biogeography   the branch of biology that deals with the geographic distribution of plants and animals.

biomass   the total quantity of living organisms of one or more species per unit of space at a given time, or of all the species in a community.

bog   an undrained or poorly drained area with vegetation of sedges and mosses (especially *Sphagnum*), often accumulating in layers of peat.

Bölling pollen zone   an interval of late glacial time, from 13,000 to 12,000 yr B.P., during which there was climatic amelioration, favoring birch and park-tundra vegetation in Europe.

brachypterous   possessing short or abbreviated wings and unable to fly.

Bronze Age   in archaeology, a cultural level (principally in Europe) from about 5500 to 3000 yr B.P., characterized by the development of bronze technology.

bucket sieve   a bucket without a bottom in which a sieve screen has been fixed near the base in order to screen large quantities of sediment.

carina   an elevated ridge or keel.

caste   in entomology, a form or kind of adult individual among social insects, such as a worker, a soldier, or a queen.

cephalothorax   the united head and thorax of arachnids (e.g., spiders and scorpions).

cluster analysis   a statistical method of grouping variables according to magnitudes and interrelationships between correlation coefficients.

clypeus   the part of the head of an insect, anterior to the frons, to which an upper lip (labrum) is attached.

cold hardiness   the capacity of an organism to tolerate cold temperatures.

coprolite   fossilized excrement.

Cordilleran Ice Sheet   an ice sheet that covered most of western Canada and the northwestern United States during the Wisconsin glaciation.

depigmentation   loss of pigment in the cuticle, often found in subterranean or cave-dwelling insects.

dichotomous key   a key to the identification of a taxonomic group in which the key characters are split into couplets.

dorsoventral compression   flattening of the body so that the dorsal and ventral surfaces are broadened and lateral regions are reduced.

Eemian interglacial   in the European stratigraphic scheme, the last interglacial.

elytron   the anterior leathery or chitinous wings of beetles, which cover the hind wings and often the abdomen.

endemic species   a species that is thought to have originated and remains only in a narrowly defined region.

eutrophic   of or perrtaining to bodies of water that are rich in mineral nutrients and organic materials and therefore productive. Oxygen may be seasonally deficient.

exoskeleton   the external skeleton of insects.

exposure   in geology, a place (either natural or man-made) where sediments or rock outcrops are exposed, such as a stream bank, a lake shore, a cliff, or an irrigation ditch.

extirpation   the extinction of a species from a given region.

extracellular freezing   freezing of interstitial liquids in the body of an organism that occurs outside the cells.

fen   a tract of low, marshy ground containing alkaline peat, rich in mineral salts, situated in the upper parts of old estuaries and surrounding freshwater lakes.

fluvial sediment   sediment laid down in running waters.

forensic entomology   the study of insects as applied to the reconstruction of crimes, especially murder.

frons   the frontal sclerite of the head capsule of insects, posterior to the clypeus.

frontoclypeus   the structure made up of the fused frons and clypeus sclerites on the head capsules of caddisfly larvae.

genotype   the genetic constitution, or total sum of the genes, of an organism.

geochronology   the reconstruction of the timing of events in geologic time, achieved through both absolute and relative dating methods.

geomorphology   the study of the origin, nature, and development of landforms.

glaciation   the covering of large land masses by glaciers or ice sheets.

Gulf Stream   an oceanic current, originating in the Gulf of Mexico and transporting warm water to northwest Europe.

halobiont   an organism that lives in saline habitats.

hemelytra   the anterior wing of bugs (Hemiptera and Homoptera); the basal portion is thickened, the apical portion is membranous.

Holocene   an epoch of the Quaternary period, spanning the interval after the last glaciation (10,000 yr B.P. to recent).

horizon   in stratigraphy, an interface indicative of a particular position in a sequence, often a distinctive, thin bed.

instars   the succession of larval stages of insects.

interglacial   a lengthy interval separating two glacial epochs in which climatic conditions were as warm as or warmer than modern conditions.

interstadial   a warm substage of a glaciation, marked by temporary retreat of ice.

Iron Age   in archaeology, a cultural level (principally in Europe) beginning about 3000 yr B.P., characterized by the development of iron technology.

isodiametric meshes   meshes of equal diameter; on insect cuticle, these are often hexagonal in shape.

kettlehole   a steep-sided basin, commonly without surface drainage, in glacial drift deposits (especially outwash), formed by the melting of a large, detached block of stagnant ice.

lacustrine deposit   sediments accumulated on the bottom of a lake.

larval exuviae   the molted skins of larvae or nymphs at metamorphosis.

late glacial   the interval of time during the waning of the last glaciation, immediately preceding the Holocene.

Laurentide ice sheet   an ice sheet covering most of eastern and central Canada and the northeastern and north-central United States during the Wisconsin glaciation.

lignite   a brownish-black coal, intermediate between peat and subbituminous coal.

littoral zone   the zone in a lake or pond extending from the shore through the depth at which plants are rooted.

macroclimate   the climate of an entire region.

macropterous   possessing large wings.

macula   a colored mark, larger than a spot.

Magdalenian   in European archaeology, a cultural period in the upper Paleolithic, at the end of the last glaciation.

mammalian megafauna   large mammal species, with adults weighing more than 44 kg.

marine transgression   the spread of seas over land areas; the spread of marine deposits over large regions previously above sea level.

marker horizon   a layer forming a sharp boundary in a stratigraphic sequence.

microclimate   the climate of a small habitat or locality.

micropaleontological card   a card, measuring 2.5 × 7.5 cm, with a rectangular cavity in which micropaleontological specimens are placed. The card is covered with a glass slide, held by an aluminum holder.

microsculpture   in entomology, microscopic sculpture, including striations, punctures, and meshes, on the surface of sclerites.

Milankovitch cycles   long-term climatic changes, which, Milutin Milankovitch (1941) theorized, are caused by changes in the earth's orbit around the sun (including eccentricity, tilt of rotation, and longitude of perihelion).

multiple regression analyses   a statistical method in which one regresses one variable on a series of other variables, taking various combinations of these to obtain a minimum of unexplained variance.

Mutual climatic range method   a statistical method of analyzing fossil insect data for paleoclimate reconstructions, based on the creation of climate envelopes for each species in an assemblage and the compilation of climatic parameters corresponding to the intersection of the climate envelopes for those species.

Neolithic   in archaeology, the youngest division of the Stone Age, characterized by the development of agriculture and animal domestication.

Oligocene   an epoch of the early Tertiary period, spanning the interval from about 33 million to 22 million years ago.

oligotrophic   of or pertaining to lakes and ponds that are low in nutrients and lack distinct stratification of dissolved oxygen in summer or winter.

orbital forcing   climatic change brought about by changes in the earth's orbit about the Sun. See **Milankovitch cycles.**

organic deposit   a sedimentary deposit rich in organic materials, such as peat or organic detritus concentrated in silts and sands.

oxidized sediments   sediments deposited in an aerobic environment, fostering decomposition of organic matter.

oxygen isotope ratio   the ratio of oxygen-18 to oxygen-16 isotopes in oxygen-bearing geologic materials (such as carbonate shells of marine organisms), used as a measure of past temperatures.

oxygen isotope stages   chronologic stages in geologic time, based upon major shifts in oxygen isotope ratios in marine invertebrate shells (see Shackleton and Opdyke, 1973).

packrat midden   well-preserved fragments of plants and animals accumulated locally by a packrat (*Neotoma* spp.) and often encased in crystallized urine.

Palearctic region   a biogeographic realm including Europe, North Africa, and northern and central Asia.

paleoecology   the study of the relationships between ancient organisms and their environments.

paleoentomology   the study of fossil insects and allied arthropod groups.

paleolimnology   the study of conditions and processes in ancient lakes.

paleosol   a buried soil horizon.

palp   a segmented tactile or tasting appendage found on the mouthparts of many arthropods, including insects and spiders.

palynology   the study of fossil and modern pollen.

periglacial environments   environments at the immediate margins of glaciers and ice sheets, greatly influenced by the cold temperature of the ice.

permafrost   permanently frozen ground, found in arctic, subarctic, and alpine regions.

phenotype   the expression of the characteristics of an organism, determined by the interaction of its genetic constitution and the environment.

phytophagous   plant-eating.

piston corer   a coring device that employs a piston inside a cylinder to reduce coring friction by creating suction.

plant macrofossils   the macroscopic remains of ancient plant fragments, including roots, stems, leaves, and fruits.

Pleistocene   an epoch of the Quaternary period, spanning the interval from 1.7 million years ago to 10,000 years ago. The Pleistocene is characterized by a series of major glaciations.

Pliocene   an epoch of the late Tertiary period, spanning the interval from about 5 million to 2 million years ago.

poikilotherm   an animal that lacks the capacity to control its body temperature.

polar front   the southern boundary of the cold, polar water mass in the North Atlantic Ocean (see Ruddiman and McIntyre, 1981).

polder   a tract of lowland reclaimed from a body of water.

Preboreal pollen zone   a term used primarily in Europe for the interval from 10,000 to 9000 yr B.P. when climate was somewhat cooler and wetter than that during the subsequent Boreal zone.

principal components analysis   a multivariate statistical technique for calculating variance along principal axes of trends in data.

proglacial lakes   lakes formed at the margins of ice sheets or glacial lobes.

pronotum   the dorsal thoracic shield of beetles and some other insects.

proxy data   in Quaternary studies, data from, e.g., fossil organisms, sediments, or ice cores used to reconstruct past environments; proxy data serve as a substitute for direct measurements of such phenomena as past temperatures, precipitation, and sea level.

pseudoscorpions   an order of small arachnids, superficially resembling scorpions but lacking stingers.

punctulae   in entomology, small punctures on the surface of insect exoskeletons.

Quaternary   The second period of the Cenozoic era, following the Tertiary and spanning the interval from about 1.7 million years ago to the present.

regression equations statistical methods of estimating the relationship of one variable to another by expressing one variable in terms of a linear (or more complex) function of the other.

riparian of or pertaining to the land bordering a stream, lake, or tidewater.

rugose wrinkled.

Sangamon in North America, the term used for the last interglacial.

sclerites in entomology, individual plates of an insect exoskeleton, separated from other plates by sutures.

seasonality the degree to which annual climatic variability is expressed in seasons.

seta in entomology, a hairlike appendage developed as an extension of the epidermal layer.

setaceous puncture a puncture on the surface of an insect exoskeleton, containing a seta.

solifluction the slow, viscous, downslope movement of waterlogged soil in regions underlain by frozen ground.

speciation collective term for the processes in evolution by which new species are formed.

stadial a substage of a glaciation, marked by glacial advances.

steppe an extensive region of dry grassland, most often used in reference to the grasslands of southwestern Asia and southeastern Europe.

steppe-tundra a mixture of steppe and tundra vegetation, developed in cold, dry conditions in Beringia during Pleistocene stadials.

stratigraphy the arrangement of strata as to geographic position and chronologic sequence.

stria in entomology, a fine, impressed line on the surface of the exoskeleton.

subalpine zone the uppermost forest zone in mountains.

supercooling the cooling of a liquid to below that liquid's freezing point without the formation of ice.

synanthropic of or pertaining to organisms that live in close association with humans and are more or less dependent on the environment created by human habitations.

synoptic-scale climate the climate over a broad area.

taphonomy a branch of paleontology, dealing with the manner of burial of plant and animal remains.

tephrochronology the collection, preparation, petrographic description, and dating of tephra or volcanic ash.

Tertiary the first period of the Cenozoic era, spanning the interval from 65 million to 2 million years ago.

thermoluminescence dating a method of dating loess and materials that have once been heated by measuring the light emitted when a sample is heated in the laboratory. Some of the energy produced by radioactive decay is stored as trapped electrons, which are released as light when heated.

thermophilous   warmth-loving; of or pertaining to animals that require warm environments to complete their life cycles.

trackway   a path made from cut tree branches and other woody vegetation laid down across moors in prehistoric times, in an attempt to create a raised surface for transportation.

treeline   the upper limit of trees in mountainous regions, or the latitudinal limit of trees in the subarctic or subantarctic regions.

trophic condition of water   the level of nutrients in a body of water, ranging from oligotrophic (nutrient-poor) to eutrophic (nutrient-rich).

unconsolidated sediment   sediment with particles not cemented together or turned to stone.

Upper Paleolithic   the youngest division of the Paleolithic, or Stone Age, characterized by the appearance of modern humans, *Homo sapiens sapiens.*

Weichselian   the last glaciation in northern Europe.

Wisconsin   the last glaciation in North America.

xeric-adapted   of or pertaining to plants or animals adapted to dry conditions.

Younger Dryas oscillation   a rapid climatic deterioration in northwest Europe, from 11,000 to 10,000 yr B.P.

INDEX